A MINIMAL METAPHYSICS FOR SCIENTIFIC PRACTICE

What are the metaphysical commitments that best 'make sense' of our scientific practice (rather than our scientific theories)? In this book, Andreas Hüttemann provides a minimal metaphysics for scientific practice, i.e., a metaphysics that refrains from postulating any structure that is explanatorily irrelevant. Hüttemann closely analyses paradigmatic aspects of scientific practice, such as prediction, explanation and manipulation, to consider the questions of *whether* and (if so) *what* metaphysical presuppositions best account for these practices. He looks at the role that scientific generalisation (laws of nature) plays in predicting, testing and explaining the behaviour of systems. He also develops a theory of causation in terms of quasi-inertial processes and interfering factors, and he proposes an account of reductive practices that makes minimal metaphysical assumptions. His book will be valuable for scholars and advanced students working in both philosophy of science and metaphysics.

ANDREAS HÜTTEMANN is Professor of Theoretical Philosophy at the University of Cologne. He is the author of *What's Wrong with Microphysicalism?* (2004) and *Ursachen* (2013), and has published many book chapters and journal articles on metaphysics, philosophy of science and early modern philosophy.

A MINIMAL METAPHYSICS FOR SCIENTIFIC PRACTICE

ANDREAS HÜTTEMANN

University of Cologne

CAMBRIDGE
UNIVERSITY PRESS

Shaftesbury Road, Cambridge CB2 8EA, United Kingdom

One Liberty Plaza, 20th Floor, New York, NY 10006, USA

477 Williamstown Road, Port Melbourne, VIC 3207, Australia

314–321, 3rd Floor, Plot 3, Splendor Forum, Jasola District Centre, New Delhi – 110025, India

103 Penang Road, #05–06/07, Visioncrest Commercial, Singapore 238467

Cambridge University Press is part of Cambridge University Press & Assessment, a department of the University of Cambridge.

We share the University's mission to contribute to society through the pursuit of education, learning and research at the highest international levels of excellence.

www.cambridge.org
Information on this title: www.cambridge.org/9781009010436

DOI: 10.1017/9781009023542

© Andreas Hüttemann 2021

First published 2021
First paperback edition 2023

A catalogue record for this publication is available from the British Library

ISBN   978-1-316-51939-4   Hardback
ISBN   978-1-009-01043-6   Paperback

Contents

List of Figures   *page* vi
Acknowledgements   vii

    Introduction   1

1 Laws of Nature and Their Modal Surface Structure   11

2 The Problem of Ceteris Paribus Clauses   38

3 Causation – Conceptual Groundwork   82

4 Causation – Application and Augmentation   108

5 Reductive Practices   129

6 Reduction and Physical Foundationalism   159

7 Reduction and Ontological Monism   186

8 Concluding Remarks: Methods and Epistemic Sources
    in Metaphysics   203

Bibliography   217
Index   229

Figures

1   Deflection of billiard ball   *page* 88
2   Transitivity I   123
3   Transitivity II   124

Acknowledgements

Large parts of this book were written during two sabbatical semesters, which were funded by the Deutsche Forschungsgemeinschaft in the context of two research units: *Causation and Explanation* (FOR 1063) and *Inductive Metaphysics* (FOR 2495). I thank the Deutsche Forschungsgemeinschaft for their support.

The research units brought together a lot of people among whom I could test out the views advocated in the following chapters. For numerous helpful discussions I am grateful to Ansgar Seide, Daniel Plenge, Oliver Scholz, Peter Hucklenbroich, Brigitte Falkenburg, Andreas Bartels, Kian Salimkhani, Kristina Engelhard, Markus Stepanians, Gerhard Schurz, Maria Sekatskaja, Dieter Birnbacher, Daniel Hommen, Alexander Gebharter, Christian Feldbacher-Escamilla, Markus Schrenk, Siegfried Jaag, Alexander Reutlinger, Marie I. Kaiser, Christian Loew and Vera Hoffmann-Kolss. The research units furthermore provided the opportunity to invite philosophers of science and metaphysicians to discuss their views in a lot of detail. The book profited a lot from these discussions and many of these found their way into this book.

I would also like to thank two anonymous referees for Cambridge University Press for their encouraging, perceptive and helpful criticism as well as Hilary Gaskin and her collaborators (Hal Churchman and Thomas Haynes) from Cambridge University Press for their encouragement and editorial guidance. Furthermore, I am indebted to Sue Clements for improving my English.

My greatest thanks go to Markus Schrenk, Christian Loew, Siegfried Jaag, Mike Hicks and Vera Hoffmann-Kolss, who read large parts of the manuscript and provided invaluable comments and help.

Some of the chapters have their origins (partly) in one of the following papers:

Hüttemann, A. (2013). A Disposition-based Process Theory of Causation. In S. Mumford and M. Tugby (eds.), *Metaphysics and Science*, Oxford: Oxford University Press, pp. 101–22.

Hüttemann, A. (2014). Ceteris Paribus Laws in Physics. *Erkenntnis*, 79, 1715–28.

Hüttemann, A. (2015). Physicalism and the Part-Whole-Relation. In T. Bigaj and C. Wüthrich (eds.), *Metaphysics in Contemporary Physics*. Poznan Studies in the Philosophy of the Sciences and the Humanities 104, pp. 32–44.

Hüttemann, A. and Love, A. C. (2016). Reduction. In P. Humphreys (ed.), *The Oxford Handbook of Philosophy of Science*. Oxford: Oxford University Press, pp. 460–84.

Hüttemann, A. (2020). Processes, Pre-emption and Further Problems. *Synthese*, 197, 1487–509.

I am grateful to the publishers as well as to my co-author Alan Love for the permission to reuse some of these materials.

Introduction

The aim of this monograph is to provide a minimal metaphysics for scientific practice. A *metaphysics for scientific practice* is the project of making explicit assumptions concerning the structure of reality that best explain the success of scientific practice. A *minimal* metaphysics for scientific practice is a metaphysics that refrains from postulating any structure that is explanatorily irrelevant for understanding scientific practice.

The main argument of the book is developed by closely analysing paradigmatic aspects of scientific practice, in particular by focusing on the questions of *whether* and (if so) *what* metaphysical presuppositions best account for these practices. More specifically, in this book, I will look at the role scientific generalisations (laws of nature) play in predicting, testing and explaining the behaviour of systems, I will analyse causal reasoning, and I will examine reductive explanatory practices. These analyses yield the following results:

(1) An account of lawhood that takes law statements as attributing dispositional properties to systems. It differs from existing accounts by (i) taking the dispositions to be determinable, multi-track properties with a complex, functional structure that have laws of composition built into them, (ii) by its insistence on systems as the bearers of these properties – in contrast to authors who deny the existence of things or systems and in contrast to authors who assume that properties are the bearers of dispositions – and (iii) by spelling out natural necessity, including dispositional modality, in terms of invariance relations. Invariance is taken to be a modal notion that accounts for all other natural modalities encountered in scientific practice.

(2) A theory of causation in terms of *quasi-inertial processes* and *interfering factors* – concepts that can be explicated in terms of the account of lawhood developed earlier. Causal dependence can be understood in terms of nomological necessity and thus in terms of invariance

relations. The account is a process theory that is able to cope with standard objections that have been raised against such theories: the problem of misconnection, the problem of disconnection and the reductionist assumption that all causation is ultimately physical.

(3) A theory of our reductive practices that makes minimal metaphysical assumptions. In particular, it rejects a foundationalist view that postulates a non-modal relation of ontological priority to account for an allegedly ontologically privileged level (e.g., as characterised by fundamental physical theories). I argue (i) that, contrary to appearances, such a view is explanatorily irrelevant for understanding our reductive practices and (ii) that arguments for a closely related eliminativist view that singles out physical facts as the only facts fails to be persuasive because they appeal to features of causation that are problematic on most accounts of causation. Scientific practice provides no reason to believe in Physical Foundationalism or Physical Eliminativism.

The rejection of an ontological hierarchy (as in (3)) suggests a new conception of what a science-informed metaphysics should look like. The positive picture that emerges is one that can be characterised in terms of 'ontological monism' and 'descriptive pluralism': It allows for a plurality of descriptions of a system (or of reality), none of which is privileged apart from pragmatic considerations. However, as descriptions of one and the same system, these different descriptions are constrained by a coherence requirement: We need to be able to understand how they fit together.

Metaphysics and Scientific Practice

My approach to the metaphysics of science distinguishes itself from other approaches in that it focuses on general features of *scientific practice* (such as prediction, explanation or manipulation) rather than on the *content* of physical theories. Many authors would agree that metaphysics (i.e., the study of the most general features of reality) needs to be informed by science. However, the details of this constraint remain controversial.

Science provides us with some of the best-confirmed claims about reality, and some of these claims seem to be directly relevant for metaphysics. So, there is a good prima facie reason to take science seriously as a source for metaphysics, as Maudlin notes:

> Metaphysics, i.e. ontology, is the most generic account of what exists, and since our knowledge of what exists in the physical world rests on empirical

> evidence, metaphysics must be informed by empirical science. As simple and
> transparent as this claim seems, it would be difficult to overestimate its
> significance for metaphysics. (Maudlin 2007, 78)

There are, however, some problems with this idea. The first concerns the
phrase 'being informed by science'. It is sufficiently vague to allow for
a number of models of the relation between science and metaphysics. This
vagueness has been widely discussed (see, for instance, Hawley 2006, Paul
2012, Ney 2012 and Ladyman 2012). I will not address this issue here
(though I will take it up in Chapter 8).

Besides the vagueness of 'being informed by science', there is a further
problem with taking the *content* of our best-confirmed theories as sources
for our metaphysics; it is generated by a version of the argument from
pessimistic meta-induction. Two of the candidates that should certainly
inform metaphysical theorising – relativity theories and quantum theory –
appear to be incompatible with each other. Theoretical physicists are still
aiming at a unified theory. While it is surely true that a potential successor
theory will reproduce the remarkable empirical successes of both theories,
it is not evident that the ontological commitments of the successor theory
will necessarily be very similar to those of the current theories. The history
of science provides no evidence that such a condition needs to be satisfied.
For instance, classical mechanics, which had a lot of empirical success that
quantum mechanics was (and is) able to account for, provided no clue to
the ontological lessons allegedly to be learned from its successor, quantum
mechanics. Thus, given that quantum theory and the general theory of
relativity are incompatible and in need of being superseded by a unified
account, it appears to be fairly risky to rely on ontological lessons directly
drawn from the content of these theories.

The point of these remarks is not that we shouldn't trust scientific
theories, or that we should completely refrain from drawing ontological
conclusions from them. The point is, rather, that we need to be aware
of the fallibility and limited reliability of this source in metaphysical
investigations.

However, I would like to propose that besides the *content* of scientific
theories there is another source that, while also fallible, might prove
particularly fruitful for the metaphysical issues under consideration in
this book. Moreover, I think that it may be a more reliable source than
the content of scientific theories. The source I'm referring to is *scientific
practice*. The scientific practice I have in mind is the practice of predicting, explaining, confirming and manipulating based on scientific

findings. In what follows I will argue that certain aspects of this practice are best accounted for by making some very general assumptions about the structure of reality. In this sense, i.e., as the explanandum in an inference to the best explanation, scientific practice can be a source for metaphysical claims.

But why should this source be more reliable than the content of scientific theories? Certainly, scientific practice in general – just like scientific theories – changes dramatically over time: Scientists constantly develop new experimental methods for observing galaxies, for probing into tissues, etc. New methods or practices of representation have been devised in mathematical physics: delta-functions have been introduced, and renormalisations group approaches enabled new ways of coarse graining in statistical mechanics. True. But from a more abstract point of view some rather general features of scientific practice, such as part-whole explanations and extrapolation, remained unchanged even when quantum mechanics superseded classical mechanics. So, one thing I will assume is that there are certain features of scientific practice that are fairly stable over a long time and can therefore be taken to be a comparatively reliable basis from which to infer metaphysical conclusions.

Although scientific practice may not be a valuable source for all metaphysical issues, I will try to show in the following chapters that with respect to the issues I am interested in, i.e., lawhood, causation and fundamentality, scientific practice is indeed a reliable and fruitful source.

The project I have sketched is *descriptive* rather than *revisionary*, to use a distinction introduced by Strawson (Strawson 1959, 9ff). But it is not *descriptive metaphysics* in Strawson's sense because it is not concerned with the structure *of our thoughts* about the world. The project is more ambitious in the sense that it is directed at what scientific practice reveals about the *structure of the world* itself. So, it is part of a descriptive, i.e., non-revisionary, but *realistically* conceived metaphysics.

But why should we assume that analysing scientific practice tells us something about the structure of reality whereas a prima facie analogous inquiry into the structure of our thoughts about the world gives us no more than very general features without which we cannot think about reality? The relevant difference is that – according to the Kantian or Strawsonian tradition – there is only *one way* we can think about reality or have objective experiences. By contrast, even though the features of scientific practice that I appeal to have proven to be fairly stable over time, there have been alternative practices of trying to cope with nature. These different practices had to prove themselves and did so with varying success.

In a letter Descartes wrote in 1638 to Morin, he appeals to a consideration of this type:

> Compare my assumptions with the assumptions of others. Compare all their *real qualities*, their *substantial forms*, their *elements* and countless other such things with my single assumption that all bodies are composed of parts. [...]. Compare the deductions I have made from my assumption – about vision, salt, winds, clouds, snow, thunder, the rainbow, and so on – with what the others have derived from their assumptions on the same topics. (Descartes 1991, 107)

Descartes' point is not only about rejecting certain ontological assumptions but also a claim about scientific practice: part-whole explanations are more successful than explanations that appeal to substantial forms, etc. Similarly, I take the very general features of scientific practice to which I will appeal to be part of a *successful* practice. By this I mean that these practices have proved successful with respect to predicting and manipulating the behaviour of the systems under investigation. Because of this success I take these practices to be good candidates for revealing something about the structure of the actual world. So, the main point is not to give a rationale or a psychological explanation of why we have a certain practice but, rather, to give the best explanation of its success, i.e., of why it works. It may, of course, be the case that the practices discussed here will be superseded by others. In this sense, the metaphysical claims advanced here are fallible.

Finally, let me mention that I am interested in *scientific* (rather than merely physical) practice. Even though most of my examples are taken from physics, the idea is that all kinds of disciplines may prima facie have a say in what is metaphysically presupposed about the world in their scientific practices. Quite a lot of science-oriented metaphysics looks for the most *fundamental physical* theories, in particular if it tries to infer metaphysics from the content of theories (North 2013, Ney 2012, French 2014, Ladyman and Ross 2007, Maudlin 2007). What is the rationale for this approach? Presumably something like the following argument: Metaphysics is the study of the most fundamental structure of reality. The sciences, too, are in the business of studying the structure of reality. Physics is most fundamental of these. Thus, what physics has to say about the world is the best starting point for doing metaphysics (see, for example, Maudlin 2003). However, as I will argue in more detail in Chapter 6, this view ultimately relies on an equivocation of the notion of *fundamentality*. Physics may very well be thought of as being fundamental in the sense of being, in principle, applicable to each and every system. Physics is the most fundamental discipline by being the most general. But it is a different question whether physics is fundamental in the sense that its subject matter is ontologically prior to the subject matter of every other discipline. Only when the latter assumption is taken for granted is it natural to assume that the theories (or practices) of fundamental physics provide the only clues for a metaphysics of science.

Overview

This book aims to give an analysis of some aspects of scientific practice and will in particular identify general assumptions about the structure of reality, which, I argue, best account for the success of our scientific practice. Most of the positive metaphysical claims are advanced in Chapters 1 and 2, while in Chapters 3 to 7 I argue that the structure outlined in Chapter 2 suffices to account for other aspects of scientific practice as well, e.g., our causal explanatory and our reductive practices. Almost no additional metaphysical structure needs to be postulated. In particular, the modal notion of an invariance relation, introduced in Chapter 1, will account for all the natural modalities and dependence relations we encounter in the scientific practice analysed in this book.

In Chapter 1, I start by analysing the role of generalisations in scientific practice. Law statements or generalisations are involved in one way or another in explanation, confirmation, manipulation or prediction. I argue that these practices require a particular reading of the generalisations involved, namely, as making claims about the behaviour of systems. These practices therefore presuppose the existence of systems or things.

Next, I look at the modal surface structure associated with laws. I use the term 'surface structure' to indicate that this structure may or may not be reduced to non-modal facts – as the Humean has it. I will sideline the debate about whether Humeanism is a tenable philosophical position. The positive claim I advance is that the modal surface structure can be explicated in terms of invariance relations – where I take invariance to be a modal notion.

In Chapter 2, I examine what appears to be a violation of the invariance of laws. Generalisations typically concern the behaviour of systems considered as isolated, while explanations, confirmations, manipulations and predictions typically concern non-isolated systems. Ceteris paribus clauses, which are often attached to law statements, take account of the fact that systems are typically not on their own. Systems are interacted on and interfered with by other systems – they are thus not invariant with respect to the behaviour of other systems. Understanding how ceteris paribus clauses work helps us to understand why we can explain, confirm or manipulate the behaviour of systems that are parts of a larger whole. An analysis of the role of ceteris paribus clauses shows that we need to read laws (generalisations) as attributing *multi-track dispositional properties* to systems. The argument relies on an analysis of scientific practice only and is not committed to more far-reaching claims, such as pan-dispositionalism, etc., that are common in the metaphysics of science literature. There is in particular no need to assume a *sui generis* conception of dispositional modality. On the contrary, I will argue that the modal aspects of dispositions can be explicated in terms of invariance relations.

In Chapters 3 and 4, I present a novel account of causation. On the one hand, it takes into consideration a certain plurality of causal concepts that has recently been diagnosed. On the other hand, it has two important features that are relevant for the overall argument of the book: (i) it shows how causation fits into a world as described by the scientific laws or generalisations, and (ii) it provides no reason to assume that causal relations are confined to a basic or fundamental level of reality – on the contrary, it might very well be the case that causation is a purely macroscopic phenomenon.

More particularly, I argue that our ordinary concept of causation ('disruptive causation') can be spelled out in terms of quasi-inertial processes and interferences. These processes and factors, in turn, can be fully explicated in terms of the generalisations provided by physics, biology or other sciences. The quasi-inertial processes in particular can be characterised in terms of *the behaviour systems are disposed to in the absence of interfering factors* and, thus, in terms of *ceteris paribus* generalisations and their underlying dispositions as discussed in Chapters 1 and 2. The analysis of the disruptive concept of causation is important for three reasons. First, this is the concept that is central for an understanding of why causal terminology plays an important role in scientific as well as everyday contexts. Second, this concept is tracked by most of our causal intuitions. I will argue that it fares much better than competing causal theories in accounting for these intuitions (this is what Chapter 4 is mainly concerned with). Third, I argue that the ordinary concept of causation is a *focal* concept: Other concepts of causation, which have been developed largely as a side effect of certain developments in the sciences and which do not explicitly rely on quasi-inertial processes and interfering factors, can be explicated with reference to this focal concept (in this sense, my explication

takes account of a certain plurality of causal concepts). One example that
will be discussed is what I call 'closed system causation': The state of
a closed system at one time causing the system at a later time to be in
a certain state.

As a result of this account of causation there is no need to make
additional metaphysical assumptions beyond those introduced in
Chapters 1 and 2. Causal relations can be fully accounted for in terms of
generalisations such as those provided by physics, biology or other sciences;
in particular, causal dependence can be explicated in terms of invariance
relations.

In Chapters 5 to 7, I will turn to reductive explanatory practices. In
the first part of Chapter 5 I distinguish various different ways in which
reductive practices are important in some sciences. I will focus in
particular on various kinds of *theory reduction* and of *explanatory
reduction*. In the second part of the chapter I argue that the rationale
for both theory reduction and explanatory reduction can be explicated
in terms of our interest in achieving an understanding of how different
descriptions of either one system (part-whole explanations) or classes of
systems (theory reduction) are related. If we are confronted with
different accounts of a class of systems (in the case of theory reduction)
or of one system (e.g., accounts of the behaviour of one system in terms
of different disciplines/vocabularies) we want these accounts to fit
together. In the case of theory reduction we want to understand why
a superseded theory was successful in the past and why we should
continue to apply it. In the case of reductive part-whole explanations
we want to understand why the description of the compound system
and those of the parts and their interactions yield the same predictions
'micro-macro coherence').

Because we consider such an understanding to be an epistemic virtue,
we seek theory-reduction and reductive explanations. I will discuss
a number of cases (quantum entanglement, phase transitions, etc.)
where we appear to have given up on reduction and coherence and
argue that this impression is misleading.

The essential point in the context of a minimal metaphysics of scientific
practice is that understanding the rationale for reductive reasoning requires
an appeal to an epistemic virtue only and does not have to rely on
metaphysical assumptions about, e.g., fundamentality or ontological
priority.

In Chapters 6 and 7, I address whether there is a different argument
for additional metaphysical structure. A quest for, e.g., micro-macro

coherence is, surely, *compatible* with further metaphysical structure. Even if the *rationale* for part-whole explanations or other reductive practices can be spelled out without metaphysical assumptions and, this does not yet tell us why this practice is successful – why it works. To answer this question, I will look in more detail at Physical Foundationalism (Chapter 6) and Physical Eliminativism (Chapter 7) because it might be that, e.g., the particular way part-whole explanations *work* requires a metaphysical underpinning, which in turn provides evidence for one of these positions.

Chapter 6 will examine what exactly is presupposed about the metaphysical character of the relation between parts and compounds in the context of part-whole explanations. These kinds of explanations have often been taken to be evidence for Physical Foundationalism, a view that assumes that an ontological priority relation obtains between the micro-level and the macro-level. I will argue that part-whole explanations (just like other explanations) do indeed presuppose the existence of dependence relations between what the explanans refers to and what the explanandum refers to (this is sometimes called a 'backing relation'). However, the stronger claim that an *ontological priority relation* obtains in nature does not do any work in understanding the dependence relations involved in our reductive explanatory practices. All we need is the assumption that parts and wholes mutually determine each other. A minimal metaphysics of science needs to postulate a dependence relation but not an ontological priority relation. Foundationalism is not implied by what classical mechanics and quantum mechanics have to say about the part-whole relation.

Nor is Physical Eliminativism – the view that the *only facts* there are, are fundamental physical facts – implied by our reductive practices, as I argue in Chapter 7. The positive picture that emerges is one that can be characterised in terms of 'ontological monism' and 'descriptive pluralism': It allows for a plurality of descriptions of a system, e.g., on a micro- and on a macro-level, none of which is ontologically privileged as the *exclusively* true account of reality, provided they are empirically adequate.

In Chapter 8, I will reflect on the methods and the character of the metaphysical arguments employed in Chapters 1 to 7. I will argue that most of the arguments can be reconstructed as inferences to the best explanation. Objections against the use of inference to the best explanation in metaphysics can be countered with respect to the specific contexts discussed in this book. What counts as the *best* explanation in these contexts can be understood in terms of what is minimally required for giving an account of the success of certain features of scientific practice. I will situate this approach by contrasting it with naturalistic and aprioristic conceptions of doing metaphysics of science.

CHAPTER I
Laws of Nature and Their Modal Surface Structure

In this and the following chapters, I advocate a practice-oriented approach to questions in the metaphysics of science. I take metaphysics to study – inter alia – the most general features of reality, among them the issues covered here: laws, causation, reduction and foundationalism. My approach starts with the *role* played by laws, causation and reduction in scientific practice. The best explanation of the success of the scientific practice we have, I argue, requires making a number of metaphysical assumptions about the structure of reality. Thus, the purpose of this and the following chapters is to examine which metaphysical assumptions we need to make in order to understand the role that laws of nature, causation and reduction play in scientific practice. In this context, our practices of explanation, confirmation, manipulation and prediction play the role of the explananda in an inference to the best explanation.

I will assume that for an explanation to qualify as the best explanation it should be *minimal*: it should contain no assumption that does not do any work in explaining scientific practice. One may worry that the conclusions we can draw from a minimality constraint thus defined depend on where we start our investigation. If we start by looking at practice P_1 and move to an investigation of P_2 and P_3, it may turn out that, given assumptions A_1 and A_2, postulating A_3 does not do any additional explanatory work. However, had we started with an analysis of P_3, A_1 might have turned out to be explanatorily irrelevant. As a rejoinder I would like to point out that our starting point is non-arbitrary. Our practice of confirmation, explanation, etc. in terms of laws of nature – as I will argue in later chapters – is presupposed in causal reasoning as well as in our reductive practices but not vice versa. Thus, we cannot avoid looking at the role laws play in confirmation, explanation, etc. outside causal or reductive contexts – and how to account for it. Analysing the role laws of nature play in scientific practice is thus the natural starting point of a minimal metaphysics of scientific practice.

We will see how far one can get with scientific practice as the main epistemic source for metaphysical arguments. Further sources that traditionally play a significant role in metaphysical theorising, such as appeal to intuitions or a preference for desert landscapes, will only be admitted if there is an argument as to why such intuitions or preferences should be considered to be truth conducive.[1]

Chapters 1 and 2 are both devoted to a characterisation of laws of nature. While the first chapter focuses on how to best reconstruct law statements and on modal aspects of laws, Chapter 2 will be concerned with the practice of hedging laws with ceteris paribus clauses. A full account of what is the best account of laws will thus have to wait until the end of Chapter 2.

I will start by arguing that the practices of explanation, confirmation, manipulation and prediction require a particular reading of the law statements[2] involved as invoking two different kinds of generalisations – internal generalisations and external generalisations (Section 1.1). Having reconstructed what law statements say in the light of how they are used in confirmation, explanation, etc., I will then explain why law statements thus reconstructed can successfully play the role they play. I argue that law statements make claims about *systems* – more precisely, attributing multi-track properties to systems (Section 1.2). Furthermore, I will analyse the modal surface structure of law statements. It is part of my approach to eschew questions as to the origin of the modal features that are delineated by laws. The question whether or not the modal surface structure is reducible to non-modal facts may be an interesting question on its own, but answers to this question typically do not do any work in explaining the success of the scientific practice we have. Still, my account will comprise three claims about the modality of laws. First, law statements attribute a space of possible states to systems. Second, laws constrain the temporal development of systems by virtue of what I will call law equations. Third, the laws' inviolability or natural necessity can be explicated in terms of the fact that they are invariant with respect to a number of different kinds of circumstances (Section 1.3).

---

[1] In Chapter 3, I argue that we do have an argument to consider causal intuitions – as opposed to intuitions of, say, law-governing or metaphysical fundamentality – to be by and large truth conducive.

[2] Although a number of people have reservations about using the expression 'law' or 'law statement' (e.g., Woodward and Hitchcock 2003, 1ff), I will use 'law statement' in order to be able to distinguish law statements from other kinds of generalisations that are involved in scientific practice (see, e.g., Section 1.1.3).

1.1 Law Statements and the Role of Different Kinds of Generalisations

1.1.1 Law Statements

Let me start with Galileo's law. It may be thought that Galileo's law is simply identical to the following equation:

$$s = {}^1/_2\, gt^2 \qquad (1.1)$$

(where s is distance covered, t is time and g is a constant). That seems wrong to me. It is fairly uncontroversial to take laws or law statements to be those (maybe complex) generalisations that play a role in extrapolation, confirmation, explanation and other aspects of scientific practice. With this characterisation of a law statement as a starting point, we can immediately infer the following consequence: if a law statement is what is confirmed or disconfirmed in trials (or used in the contexts of explanation, prediction or manipulation), an equation on its own cannot be an example of a law (or a law statement – in what follows I will use these two terms synonymously). As a matter of fact, nobody takes Galileo's law to be disconfirmed by balls uniformly rolling on a horizontal plane or by stones lying on the ground, both of which fail to satisfy Eq. (1.1). What is missing is a claim about *the kinds of systems* that are meant to be represented by the equation. Galileo's law is not simply a mathematical equation. Nor does it suffice to add that t represents time and s the path taken by an arbitrary object. Galileo's law is the claim that the behaviour of a particular class of systems can be represented by this equation. A full statement of Galileo's law might thus be something like the following:

Free-falling bodies behave according to the equation $s = {}^1/_2\, gt^2$.

Similarly, $F = ma$ is merely a mathematical equation. It becomes a law statement once it is asserted that this equation is meant to represent the behaviour of physical systems; indeed, of all physical systems whatsoever. And again, the Schrödinger equation with the Coulomb potential on its own does not qualify as a law statement; that is, it is not what we confirm or disconfirm. By contrast, the claim '*Hydrogen atoms* behave according to the Schrödinger equation with the Coulomb potential' is a law statement.

The fact that equations such as $s = {}^1/_2\, gt^2$ come with a domain of systems[3] for which they are meant to be relevant has been noted by others, e.g.,

---

[3] Nothing hinges on the term 'system' – 'object' or 'thing' would be fine too. I say a little bit more about systems in Section 1.2.

within the semantic account of theories. Thus, Bas van Fraassen, referring to Ronald Giere, defines a *theory* (not a law[4]) as consisting of:

(a)   the *theoretical definition*, which defines a certain class of systems; and
(b)   a *theoretical hypothesis*, which asserts that certain (sorts of) real systems are among (or related in some way to) members of that class (van Fraassen 1989, 222).

A preliminary general characterisation of law statements might thus be the following:

(A) All systems of a certain kind K behave according to Σ.

Here 'Σ' – the law predicate – typically stands for an equation or a set of equations. The expression 'of a certain kind K' may refer to all physical systems whatsoever, as it does in the case of Newton's second law or in the case of the bare Schrödinger equation. Or it might refer to a more circumscribed class of systems such as free-falling bodies or hydrogen atoms, thus giving rise to so-called *system laws*.[5] It is important to note that the behaviour attributed to the systems in question is in general complex and relational. In the case of free-falling bodies, the length of the path and the time taken are related, not only for actual values of the variables but for all possible (or some restricted domain of) values. Taking 'All ravens are black' as a paradigm for law statements ignores the complex structure usually attributed to systems.

Another example that illustrates the structure of law statements is Euclidean geometry. Euclidean geometry on its own is a mathematical theory without any empirical import. We get an empirically testable claim (a law) if we take a certain class of systems (space-times) to be adequately characterised in terms of Euclidean geometry.[6]

---

[4]   In fact, Giere and van Fraassen deny that there are laws of nature (van Fraassen 1989, 183ff). According to my reconstruction, what Giere and van Fraassen call a 'theory' should be taken to be a law statement.

[5]   One might worry about the exact characterisation of the system to which Σ is attributed. The worry is that one needs Σ to individuate the systems in question. That, of course, would make the law statement an analytical truth and thus devoid of empirical content. It has to be assumed that the relevant class has been individuated antecedently, for example in terms of experimental procedures ('free falling bodies') or by other means that do not depend on Σ. This is a thorny issue that I will not go into in this book.

[6]   This point was famously observed by Einstein: 'As far as the laws of mathematics refer to reality, they are not certain; and as far as they are certain, they do not refer to reality.' In fact, Einstein – in his paper 'Geometry and Experience' (Einstein 1921) – suggests a view of laws or theories pretty close to the one suggested here.

Law statements as just characterised play a prominent role not only in physics but also in other disciplines. Thus, the Lotka–Volterra equations describe the temporal development of a biological system consisting of two populations of different species, one being a predator and the other prey. The relevant equations for prey and predator populations are (1) $dx/dt = x(a - by)$ and (2) $dy/dt = -y(c - gx)$, where x represents the number of prey and y the number of predators and a, b, c and g are constants. Again, we can distinguish between the system to which the equations apply, on the one hand, and the equations or the description of the behaviour, on the other.

Even in cases in which the behaviour in question is not represented mathematically, it is possible to distinguish the behaviour from the systems to which it is attributed. Thus, according to Schmalhausen's law, a population at the extreme limit of its tolerance in any one aspect is more vulnerable to small differences in any other aspect. On the one hand, we have populations (the systems); on the other, we have a qualitative description of what may happen to the population (the behaviour).

1.1.2   Internal and External Generalisations

Characterising law statements in terms of (A) allows me to draw attention to an important distinction between different kinds of generalisations. Take the example of Galileo's law,

Free-falling bodies behave according to the equation $s = \frac{1}{2} gt^2$.

Even though there are often no explicit quantifiers, law statements usually involve at least two different kinds of generalisations (as will be illustrated by an example in Section 1.1.3). In Galileo's law we can distinguish one form of generalisation that quantifies over systems (for all x that are falling bodies). This quantification specifies the *objects* (or systems) to which a certain kind of behaviour is attributed. Besides generalisations that pertain to objects or systems, there are *system-internal generalisations*. These generalisations concern the values of the variables that appear in the equation. When we claim that a system behaves according to the equation $s = \frac{1}{2} gt^2$, what is implied is that for every value of t the path s that the body has fallen is determined by $s = \frac{1}{2} gt^2$.

We can therefore distinguish two kinds of generalisations (see Scheibe 1991a)[7]:

---

[7] Hitchcock and Woodward (2003, 189) draw attention to this distinction, albeit in different terms, when they remark with respect to explanation that 'the nomothetic approach has focused on

(1)   *System-internal generalisations*: Generalisations concerning the values
      of variables. For instance, in the case of Galileo's law, the system-
      internal generalisation is that the equation holds for all values of the
      variable t (or at least for all values within a certain range).

(2)   *System-external generalisations*: Generalisations concerning different
      systems such that the equation pertains to all systems of a certain kind
      (e.g., free-falling bodies).

In the case of the Lotka–Volterra equations, the internal generalisations
concern the variables x (number of prey) and y (number of predators),
while the external generalisations concern ecological systems consisting of
prey and predator populations.

In our preliminary law characterisation (A), the system-external generalisation ('All systems of a certain kind') is explicitly mentioned while the
system-internal generalisations are implicit in Σ. One reason why internal
generalisations are not made explicit may be the fact that usually more than
one internal generalisation is allowed by the law equation and it is the
context that determines which of those are relevant for the characterisation
of a particular phenomenon. To be a bit more specific, law equations (in
contrast to the structural equations discussed in the causation literature)
are not in general asymmetric. As a consequence, a law equation, such as
$pV = vRT$, allows us to infer not only that once the values for p and V are
given those of T are determined but also that the values for p and
T determine those for V, etc. The law equation thus implies at least three
different internal generalisations. Which of those is relevant for a particular
situation may depend on the quantities on which we want to intervene or
on other features determined by the context. Similarly, when we claim that
a system behaves according to the equation $s = {}^1\!/_2\, gt^2$, what is implied is not
only that for every value of t the path s is determined by $s = {}^1\!/_2\, gt^2$ but also
that for every path s, the time t that the body has taken to fall is determined
by $t = \sqrt{(2s/g)}$. The law statement allows us to assert both generalisations.

The distinction between internal and external generalisations goes
hand in hand with a distinction between different kinds of counterfactuals
noted by Woodward and Hitchcock. First, we have what Woodward and
Hitchcock (2003, 20) call 'other object counterfactuals'. Examples of other
object counterfactuals include 'If b had been a raven it would have been
black' or 'If b had been an ideal gas, it would have behaved in accordance

a particular kind of generality: generality with respect to objects or systems other than the one whose
properties are being explained'. By contrast, their own account of explanation relies on generalisations that pertain to the values of variables.

with the equation $pV = v\,RT$.' These counterfactuals presuppose external generalisations; traditional accounts of laws as external generalisations stress the fact that laws support other object counterfactuals.[8]

Woodward and Hitchcock contrast other object counterfactuals with *same object counterfactuals*. These pertain to particular systems, such as in the statement 'If the ideal gas in question had had a volume $V = V_o$ and a pressure $p = p_o$ its temperature would have been $T = T_o$.' Same object counterfactuals presuppose internal generalisations. Woodward and Hitchcock argue that it is same object counterfactuals that are relevant for scientific explanation. I largely agree, though I do not accept the interventionist account of the truth conditions for these counterfactuals. In Chapter 2 (Section 2.4.3), we will encounter a third kind of counterfactual connected with law statements.

The fact that law statements come with internal generalisations is a first hint at the complexity of what law statements assert. Take the ideal gas law as an example. The law statement is highly complex because it is a functional law. It implies an infinite number of statements of the form 'If the value of p and the value of V of a particular gas had been such and such then the temperature T would have been thus and so.' Thus, the behaviour or properties that law statements attribute to systems are typically infinitely *multi-track* (I will examine this issue in more detail in Section 2.4.3). Note that the infinity of implied statements does not preclude that the ideal gas law can be stated in finite terms.

The Schrödinger equation provides another illustration of the complexity of law statements. When we claim that hydrogen atoms can be characterised in terms of the Schrödinger equation with the Coulomb potential, the Σ in our canonical statement 'All systems of a certain kind K behave according to Σ' comprises the conceptual apparatus of quantum mechanics. So, when we say that hydrogen atoms behave according to the Schrödinger equation with the Coulomb potential, we are saying that they behave according to quantum mechanics in which the Schrödinger equation is concretised via the Coulomb potential. The essential point is that law statements attribute a behaviour to systems already identified as being of a certain kind by invoking law predicates that typically involve a highly complex mathematical apparatus. This complexity becomes invisible if we are operating with examples of the 'All ravens are black' sort.

---

[8] The role of other object counterfactuals is somewhat controversial because they may be thought to involve metaphysical impossibilities in the antecedent (see Tan 2019 for discussion). That debate, however, is not relevant for the purposes of this chapter.

Examples of the latter kind are misleading because they suggest that law statements can be analysed solely on the basis of external generalisations. This assumption shaped much of the debate about laws of nature in the twentieth century.

1.1.3   Excursus: The Role of Internal and External Generalisations in Standard Explanation

The distinction between internal and external generalisations will prove fruitful in later sections of this chapter. It also helps to understand how certain simple cases of explanations work. Consider an example of what may be called a 'standard explanation'(for a discussion of this example see Skow 2016, 75ff.).

An explanation or answer to the question 'Why did that rock hit the ground at a speed of 4.4 m/s?' might consist in the statement 'It hit the ground at that speed because it was dropped from a height of one metre.' If the further question arises as to why the one explains the other, an answer will refer to the equation $s = \frac{1}{2} gt^2$.

How exactly does this explanation work? The explanandum in this case is the velocity of a particular rock (more generally, the state of a system). The explanans mentions the height and the equation $s = \frac{1}{2} gt^2$. So, we have the general structure that in a standard explanation we explain the state of a system in terms of initial conditions and a (dynamic) law equation.

The explanation points out how the speed of the rock – in virtue of the equation $s = \frac{1}{2} gt^2$ – depends on the height from which the stone was dropped; in doing so, the explanation appeals to an internal generalisation. By contrast, the fact that Galileo's law *covers* the rock, as implied by the external generalisation, is a *presupposition* for the explanation to work. Without knowing that the law covers the rock it would not make sense to explain the velocity in terms of this equation.

More generally, in simple standard explanations like the one just described, the external generalisation claims that the law predicate Σ is relevant for the explanandum. This appeal to an external generalisation is not a part of the explanans but rather, a presupposition of the explanation. By contrast, internal generalisations typically figure explicitly in the explanans since they provide information about how one quantity depends on other quantities.[9]

---

[9] How does this account relate to Skow's distinction between levels of reasons why? Skow (2016, chapter 4) distinguishes a first-level question 'Why did the rock hit the ground at 4.4\m/s?' (Answer/

In the 1960s, Scriven and Hempel debated the role of laws in explanation. One of the contentious issues was whether laws explicitly figure in the explanans or whether they provide a 'role justifying ground' 'roughly showing that the explanans is relevant for the explanandum' (Salmon 1984, 17 fn6). By relying on the distinction between external and internal generalisations we see that law statements play both roles. The external generalisation does not figure explicitly in our simple standard explanation; it provides a 'role justifying ground'. By contrast, the internal generalisation is explicitly appealed to in the explanation.

To sum up: what I do hope to have shown is that not only external but also internal generalisations can play a role in explanation and that the distinction is thus significant for understanding scientific practice.

1.2   Systems

Up to this point I have reconstructed what we should assume law statements to say given their role in confirmation, explanation, etc. I will now turn to an explanation of why law statements thus reconstructed can successfully play the role they play.

In Section 1.1.1, I argued that law statements attribute a certain kind of behaviour to systems: All physical systems of a certain kind K behave according to Σ'? A pretty straightforward explanation of why we can successfully work with such law statements is the assumption that systems that display the relevant behaviour do in fact exist.

To talk of 'systems' and 'behaviours' is very common in the sciences. Systems come in all kinds of sizes, including space-times, economies, interacting predator–prey populations, cells, gases and hydrogen atoms. In contrast to the notion of substance (on some interpretations), systems do not carry with them notions of fundamentality or indivisibility; systems

---

reason: it was dropped from 1 m) from a second-level question 'Why is being dropped from 1\m a reason for it having the speed of 4.4\m/s?' (Answer/reason: $s = \frac{1}{2} gt^2$, etc.) I have not made this distinction between these different levels of why questions, but it would do no harm to my argument if I did. What is important for me is to distinguish a third-level question that Skow does not consider: 'Why is it that the equation $s = \frac{1}{2} gt^2$ is relevant for the second-level answer?' An answer to the third-level question has to appeal to the external generalisation, while an answer to the second-level question appeals to the internal generalisation. Thus, for my purposes the issue of whether first and second level need to be distinguished is not relevant; what is relevant, however, is that the third level can be distinguished.

may be fairly complex, they might be constituted out of subsystems, and so on.[10]

In the case of physical systems, we can distinguish different aspects of their behaviour. Some quantities of a physical system are constant; others vary with time. For instance, for a single classical particle we can distinguish position and momentum as changing quantities, whereas mass remains constant. The values of the varying quantities at a particular time are called the 'state' of the physical system at this time. However, the constants and the state of a system do not determine the complete behaviour of the system. We also have equations that describe the connections between the various quantities involved and in particular how the state of the system develops over time: its *temporal development* or *dynamics*.

Talk of 'behaviour' indicates that often what is attributed to systems is not just a set of static properties but rather a certain temporal development. The Lotka–Volterra equations, for example, tell us how the predator–prey system will develop over time. Similarly, the Schrödinger equation applied to certain kinds of systems describes their dynamics.

I have argued that, in order to make sense of the fact that law statements are taken to be confirmable or disconfirmable, we need to understand them as attributing behaviour to systems. The assumption that we can identify systems that behave in certain ways is thus essential for understanding scientific practice, at least to the extent that this practice involves law statements.

Thus, whatever the content of a scientific theory or law, that is, whatever structure is attributed to reality by the law predicate Σ, the analysis of this structure cannot show that there are no systems or that there are no things. For instance, it is sometimes suggested that contemporary physics, particularly the phenomenon of indistinguishable particles, shows that there is no place for objects (or systems) with intrinsic natures in metaphysics (Ladyman and Ross 2007, 131). Such a claim is a non-sequitur, as I will now argue.

For the purpose of illustration, consider a two-electron system. (Normalised) vectors in two-dimensional Hilbert spaces represent the spin states of the separate particles. The possible spin states of the

---

[10] As I will argue in more detail in Chapter 3, nature may suggest but does not dictate how to draw the boundaries of the systems in which we are interested; that is, it does not dictate how to individuate systems. This is particularly obvious when we are dealing with macroscopic systems. A commitment to systems is thus not a commitment to nature having joints that are completely independent of pragmatic considerations (see Section 3.5.2).

compound system are all those states that can be represented as (normalised) vectors in the tensor product of the Hilbert spaces associated with the separate particles $H_s = H_1 \otimes H_2$. If we take as a basis for H_1 the eigenvectors in the spin z-direction $|\psi^{z\text{-}up_1}\rangle$ and $|\psi^{z\text{-}down_1}\rangle$ and as a basis for H_2 $|\psi^{z\text{-}up_2}\rangle$ and $|\psi^{z\text{-}down_2}\rangle$, the following superposition state will be among the possible states of the (normalised) compound system:

$$\Phi = 1/\sqrt{2}\, |\psi^{z\text{-}up_1}\rangle \otimes |\psi^{z\text{-}down_2}\rangle - 1/\sqrt{2}\, |\psi^{z\text{-}down_1}\rangle \otimes |\psi^{z\text{-}up_2}\rangle.$$

If the electrons are in such an entangled state, it is true that the electrons cannot be described as two individual particles with intrinsic properties, where intrinsic properties are conceived of as properties that the systems have independently of the properties of other systems (see, e.g., Ladyman and Ross 2007, 135ff). However, even if the indistinguishability and non-individuality of the electrons are granted, it does not follow that there are no things or systems. The system to which the entangled state is attributed (the 'two-electron system') may still be a system with intrinsic properties or states.

The problem with Ladyman and Ross's claim is not only that the argument from the indistinguishability of particles to the claim that there are no things or systems is a non-sequitur. In fact, when we try to confirm claims about the indistinguishability of particles, we have to presuppose the existence of things or systems.

This fact is nicely illustrated by the research that was awarded the Nobel Prize in Physics in 2001. As one of the recipients of the prize notes, 'The phenomenon of Bose-Einstein condensation (BEC) is the most dramatic consequence of the quantum statistics that arise from the indistinguishability of particles' (Ketterle 2007, 159). The prize was awarded for the empirical confirmation of the consequences of the indistinguishability of particles, specifically 'for the achievement of Bose-Einstein condensation in dilute gases of alkali atoms, and for early fundamental studies of the properties of the condensates' (Nobel Prize press release 2001). What this reinforces is that confirming the empirical consequences of the indistinguishability of particles requires systems or objects, such as 'dilute gases' or 'condensates'.

Furthermore, the Nobel Prize-winning research concerning the indistinguishability of the parts of a system does not tell us that the properties of the condensate fail to be intrinsic. Bose–Einstein condensates may very well be systems with intrinsic properties or behaviour, which can be studied independently of relations to other systems. It tends to be overlooked that law statements are *statements about systems* if theories are simply taken to be sets of equations and if they are analysed without taking into account that their empirical import is generated by the assumption that these are true of something in the real world.

It may be objected that considerations of quantum entanglement lead us to the view that there is only one system with intrinsic properties – the universe as a whole. That may very well be true, but it would still be in conflict with the claim that there are *no* systems. Furthermore, what we need to understand is why we can successfully treat subsystems of the universe as if they exhibited their behaviour independently of other systems in the world (more on this in Chapter 2).

According to another objection, there is less of a conflict than it first appears between Ladyman and Ross's denial of systems or things and my insistence that we have to assume their existence. While it is true that structural realists sometimes claim that there are *no things* (Ladyman and Ross 2007, p. 130: 'a first approximation to our metaphysics is: "There are no things. Structure is all there is"'; see also French 2014, chapter 7: 'The Elimination of Objects'), neither Ladyman and Ross nor French deny that we can meaningfully talk about (everyday) objects. Both have their ways of accommodating true assertions about medium-sized objects in a world that (according to their view) strictly speaking does not contain any. Thus, according to French, 'we can reject tables, people, everyday objects in general as elements of our fundamental ontology, whilst continuing to assert truths about them' (2014, 167). However, in contrast to Ladyman, Ross and French, for the reasons outlined in this section I don't think we can reject systems or objects as elements of our fundamental ontology. They cannot be analysed away if we want to understand the success of our scientific practice.

Let us now have a closer look at how Ladyman and Ross argue against intrinsic natures:

> [. . .] talk of unknowable intrinsic natures and individuals is idle and has no justified place in metaphysics. This is the sense in which our view is eliminative; there are objects in our metaphysics but they have been purged of their intrinsic natures, identity and individuality, and they are not metaphysically fundamental. (Ladyman and Ross 2007, 131)

It seems, however, that scientific practice in at least some cases gives us very good reasons to suppose that there are systems with *intrinsic* properties, provided an intrinsic property can be taken to be a property of a system that the system has independently of the properties of other systems and its interactions with such systems.

In experimentation we typically try to shield from interfering factors; we try to causally isolate the system under investigation – that is, we try to figure out how the system would behave if it were on its own – not interacting with the world. Shielding and isolation point to the fact that in experimentation we try to determine properties of systems that these systems have independently of the properties of other systems and their interactions with such systems. Presumably, even in the case of the experimental determinations of the behaviour of dilute gases of alkali atoms, as well as studies of the properties of the condensates, the experimenters determined intrinsic properties of the systems in question.

If we take into consideration scientific practice as a source of metaphysics of science, we have very good reasons to stick with at least some of the features Ladyman and Ross classify as 'standard metaphysics' (Ladyman and Ross 2007, 151) – namely, that science deals with systems and their intrinsic properties.

1.3 Modal Surface Structure

What I have argued so far is that law statements attribute complex (multi-track) behaviour to systems via law predicates. I will now examine the modal aspects of this behaviour in more detail. Traditionally, when the nomological or modal character of laws is discussed, the focus is on external generalisations of the kind 'All Fs are Gs.' Armstrong, for instance, argues that the properties of, say, being an electron and having a certain charge are related by a *sui generis* relation of nomological necessitation that explains why all electrons have charges. Bird holds that the fact that all negative charges repel each other obtains in virtue of the negative charges' essence. In this section I will attempt to show that examining internal generalisations and their role in scientific practice will unfold a rich modal structure underlying the characterisation of the behaviour of systems. More particularly, I will advocate the following claims: Law statements attribute a space of possible states to systems (Section 1.3.1). Laws constrain the temporal development of systems by virtue of law equations (Section 1.3.2). The laws' ability to constrain, their natural necessity, can be explicated in terms of the fact that they are invariant with respect to a number of different kinds of circumstances (Section 1.3.3).

I use the term 'modal *surface* structure' because even though I will argue that nomological or natural necessity, as well as all the natural dependence relations we will encounter in later chapters (dispositional modality, causal dependence, part-whole dependence), can be explicated in terms of invariance, invariance is a modal notion itself, and I will make no attempt to reduce modal facts to non-modal facts.

1.3.1   A Space of Possibilities

In order to understand the role of internal generalisations in scientific practice we need to distinguish two features. First, internal generalisations quantify over *a domain of values for variables*, which serve as possible initial or boundary conditions. Second, internal generalisations typically include a *law equation*. The law equation restricts or determines the values for the variables of the system or the values for variables that characterise the temporal development of the states of the system. I will deal with the law equation in the next section.

By virtue of internal generalisations, laws attribute a space of possible states to systems. In the case of dynamical laws, it is assumed that the systems have a set of possible initial states. With respect to these states we can distinguish two cases. Either the domain of quantification comprises all possible states (e.g., in the case of Newton's second law or the Schrödinger-equation) or, as is the case in more specific laws, the domain of quantification comprises only a restricted range of states. Hooke's law, for example, holds only for a limited range of elongations.

What is essential for our investigation is the fact that in both cases we are dealing with a *modal* presupposition because it is not only actual states or actual behaviour with which the internal generalisations are concerned. In fact, the internal generalisation on its own does not even tell us which state of the system is the actual state. The internal generalisation's concern is possible behaviour only (whether actual or non-actual). Thus, the fact that internal generalisations come with a domain of values for variables requires the assumption that law statements attribute a *space of possible (and mutually exclusive) states* to systems.

That laws attribute to systems a space of possible behaviour has recently been discussed as a threat to Humean accounts of laws of nature, because (a) it seems to be prima facie problematic to square this feature with the requirement of informational strength and (b) it may indicate that laws give not only information about patterns in the Humean mosaic but also genuinely modal information (see Hall 2015; Hicks 2018; Jaag and Loew 2020). For instance, Ned Hall observes that

it is worth noting that breadth or permissiveness of the [range of initial conditions] makes for a certain kind of explanatory strength. For it is, other things equal, a point in favor of a physical theory that it recognizes a wide range of nomologically possible initial conditions. Compare, for example, Keplerian and Newtonian accounts of the solar system. Granted that the Newtonian account is much more empirically accurate; it is also, from the standpoint of scientific investigation, better in a distinct sense: for it allows us to answer questions not merely about how the elements of the solar system did, do, and will behave, but also about how they would have behaved under alternative physical conditions. (Hall 2015, 263)

Many laws of nature – in particular the dynamical laws of fundamental physical theories – allow for a wide range of initial conditions. The fact that laws come with a range of possible states is essential for the role laws of nature play in scientific practice, as the following examples illustrate.

One case is the application of law statements in engineering contexts. Suppose an engineer considers different ways to build a bridge. Specifically, she will be considering different (e.g., Newtonian) models for the bridge: she will consider a scenario S_1 in which the bridge is built with materials M_1 and a scenario S_2 in which the bridge is built with materials M_2. First, the engineer will determine how Newtonian mechanics describes what would be the case if the bridges were built. She will furthermore ascertain what would happen in these models if certain parameters were varied: whether the hypothetical bridge would remain stable if the traffic were of a certain kind, if the weather conditions changed, and so on. Hence, a scientist or engineer will be interested not only in what is actually the case, but also in what is non-actual but (nomologically) possible. Such information about what is possible and what isn't is needed in order to know how to manipulate a system such that it reaches a designated state, for instance that the bridge doesn't collapse given the expected traffic.

More generally, it might be argued that laws play a role in decision making. It is constitutive for decision making that various possible outcomes are examined – in order to explore how different situations would develop – due to the laws of nature. Laws can only play this role because of the fact that they come with a range of (possible) initial conditions.

Furthermore, in the explanation of events – if we follow Woodward and Hitchcock's account of explanation – we appeal to laws or generalisations not because they tell us what is actually the case but because they provide modal information. According to their account, we can explain why a gas G has a certain temperature T_0 by showing how the temperature

T depends on the pressure p of the gas and its volume V (Hitchcock and Woodward 2003, 4). Hitchcock and Woodward contrast their account with the deductive nomological account of explanation:

> the generalization [...] not only shows that the explanandum was to be expected, given the initial conditions that actually obtained, but it can also be used to show how this explanandum would change if these initial and boundary conditions were to change in various ways. (Hitchcock and Woodward 2003, 4)

According to Woodward and Hitchcock, the counterfactuals that are explanatory do not only appeal to information about what is actually the case (or what was the case) but also to nomologically possible but non-actual behaviour of systems. This possible but non-actual behaviour is characterised in terms of the law equation and the domain of quantification of the internal generalisation. The counterfactuals rely on non-actual but nomologically possible states of the gas and thus on the modal structure that the law statements attribute to systems.

To conclude: Internal generalisations on their own do not tell us which state a system is actually in. To determine the actual state of the system we need additional information – information about the actual values of the variables that characterise the system. With respect to the states of a system, the law statement (by virtue of the domain of internal generalisations) gives us *purely modal* information: information about (nomologically) possible and mutually exclusive states in which the system might be.

1.3.2 Constraints

In the previous section I argued that law statements attribute a space of possibilities to systems due to the fact that internal generalisations quantify over a domain of values for variables that represent mutually exclusive possible states of a system. Let me now turn to the second aspect of internal generalisations that is relevant for the examination of modal structure, the law equation. I will argue that we need to assume a further modal feature to make sense of our scientific practice concerning law statements: law statements do not simply register the past, present and future behaviour of systems; they describe how this behaviour is constrained. While in this section I will introduce the claim, in the next section I will argue that if we understand this claim in terms of invariance relations, this best explains a certain feature of scientific practice, namely, why we can rely on laws.

Internal generalisations put *restrictions* on the space of possible behaviour of systems by establishing relations between variables (i.e., law equations). These restrictions can either concern the synchronic co-possibility of values of variables that characterise the state of a system – as in the case of the ideal gas law – or the temporal evolution of the states of a system – as in the case of the Schrödinger equation.

In the case of a synchronic law (law of coexistence), such as, e.g., the ideal gas law, the set of possible values for the variables p, V and T is restricted to those that satisfy the equation $pV = \nu RT$. Thus, the possible states of the gas are constrained to a two-dimensional hypersurface of the three-dimensional space that is generated by the variables p, V and T. The internal generalisation does not only provide information about how the *actual state* of a system (if known) is constrained. In addition, it tells us how all possible states of the system are constrained, whether or not they are actual. That the systems are constrained means that those states not on the hypersurface are not accessible to the system. They are classified as states the system cannot possibly occupy, given the law equation, i.e., as nomologically impossible states.

The fact that the gas satisfies the equation of the gas law allows a scientist or an engineer who is able to manipulate pressure and volume to ensure that the gas will have a certain temperature. Similarly, the engineer might want to prevent certain situations, such as preventing a gas from having a certain temperature. In such cases she will rely on the fact that the law tells us that certain combinations of pressure, volume and temperature will not occur; by setting pressure and volume appropriately we can make sure that a certain temperature value will not obtain.

The same holds for internal generalisations that describe the temporal evolution of a state of a system. Provided we prepare the system under consideration in a certain state, and provided the equation in question is deterministic, we can ensure that at a later time the system is in a certain state, and we can also prevent the system from being in certain other states.

In the case of prevention, it is not only that given certain combinations of, say, pressure and volume, certain values for T simply do not occur. There is a sense in which these values *cannot* occur.[11] The use scientists and engineers make of internal generalisations in scientific practice is best

---

[11] For this reason, Popper conceived of laws as 'prohibitions' (Popper 1959, §15).

understood by assuming that internal generalisations represent modal, that is, nomologically necessary, relations.[12]

That internal generalisations ought to be understood in this way requires taking a certain perspective, the perspective of a scientist or engineer operating *within* the universe, in contrast to the omniscient outside observer whose sole job is to document the world. In discussions about laws of nature the dominant perspective is that of 'a scientist operating outside the universe and looking in. This ideal scientist starts with the knowledge of all the facts of the world, so the only task left to her is to organize them' (Hicks 2018). This ideal, inward-looking scientist has two pertinent features: she is omniscient and she is interested in the description or the organisation of facts only. This, however, is not the perspective taken in scientific practice and it leaves out why laws are best understood as – at least prima facie – representing nomologically necessary relations. The essential difference is that a real engineer or scientist – in contrast to the ideal scientist – will *rely* on the internal generalization in the sense that the generalization tells her that any other value than the one that is determined by the equation $pV = v\,RT$ *cannot* occur.[13] The notion of invariance that I introduce in the next section will clarify what it means that a scientist relies on laws.

1.3.3   Invariance

It may seem that in the previous section I have illegitimately smuggled in modal terminology. Instead of saying that the law equation of an internal generalisation states a relation between the values of different variables, I have said that the values of the variables are *constrained* or *restricted*, that they obtain with *nomological necessity*. Why this modal terminology? Why claim that a certain value of T *cannot* (fail to) occur? Why argue that, provided certain values for p and V, the occurrence of a certain value of T is *nomologically necessary* and that we can rely on this?

The essential point is that laws – as opposed to accidental generalisations – do not simply state that a relation between variables obtains. Looking at

---

[12] As a reminder, I am not arguing against the Humean at this point; I am interested in the modal *surface structure* that may or may not be reducible to non-modal facts.

[13] A focus on laws as instruments or tools for limited beings may have always been part of Lewis's best system analysis (Jaag and Loew 2020). Recently, there have been various attempts to argue along these lines, that is, to claim that laws should be conceived as instruments for cognitively limited beings that operate inside the universe (Hicks 2018; Jaag and Loew 2020; Ismael 2015). However, in the past, in the literature on laws of nature there has always been a focus on explanation and description rather than on manipulation.

scientific practice, i.e., not simply looking at the law statement or the law equation itself but examining their role, i.e., how they are used, reveals that law statements should be understood as implying independence or invariance claims. For example, the ideal gas law is understood as implying that in a gas the value for T is fixed *under a wide variety of circumstances*: if certain values for p and V are fixed, then *whatever other features the gas may have and whatever else is going on in the universe*, a certain value for T is determined. It is part of how the content of the internal generalisation is understood that it is only the values of p and V that determine the value for T. An essential aspect of what laws tell us is that those variables that *do not* occur in the law equation are irrelevant for the determination of the values of certain other variables whatever the circumstances may be.

The ideal gas law, if true, is not a mere truth. Its being a law means that nobody could bring about a situation such that it is false. No person or government could bring it about that in an ideal gas the values of pressure, temperature and volume fail to be on the two-dimensional hypersurface that is determined by the ideal gas law.[14] In this sense, the behaviour of the gas is constrained or nomologically necessary. By contrast, an accidental generalisation may state that a certain equation or correlation between, say, variables representing the bread prices in London and the water levels in Venice may actually obtain. However, the equation representing the accidental generalisation is not taken to be invariant – it is not taken to continue to hold if, for instance, the British government fixes the price of bread.

The fact that the law and the accidental generalisation are used differently, that they play a different role in scientific practice, is best explained by assuming that laws state relations that are invariant. An illustration of this claim is that we can rely on laws but not on accidental generalisations, precisely because the former imply, as I argued before, that many features of the universe are irrelevant for the determination of the behaviour of the system we are interested in. The engineer or scientist *in* our world – in contrast to the ideal, inward-looking scientist – does not know all the facts of the world. She is confronted with epistemic risk when she is predicting, manipulating or constructing systems. What we need to explain is why this scientist or engineer can rely on what she takes to be laws. What is it about laws that accounts for the possibility of relying on them? With the notion of invariance, we can make sense of

[14] The same holds for real gases and more realistic gas law equations such as the Peng–Robinson equation.

why the scientist and the engineer rely on (what they take to be) laws rather than on accidental regularities.

Suppose a scientist S at a certain time has encountered the same number of positive instances for two claims, P_1: $\forall x \ (Fx \rightarrow Gx)$ and P_2: $\forall x \ (Mx \rightarrow Nx)$. Suppose that in the case of P_2 but not in the case of P_1 there is furthermore evidence for invariance – evidence, say, that P_2 holds under all kinds of changes of the behaviour of other (e.g., neighbouring) systems. For S it is now much more reasonable to be confident that P_2 will continue to hold, compared with P_1, because there is evidence for the fact that P_2 is stable and will not break down if changes in the environment occur. S can and will rely on P_2 much more so than on P_1. Thus, if having evidence for the nomic character of a generalisation means – as I have argued – having evidence for invariance relations, we can understand why scientists and engineers rely on laws much more than on other generalisations.

To sum up, the role that internal generalisations and in particular law equations play in scientific practice is best understood by assuming not only that they describe relations between variables but also that they imply that these relations hold with nomological necessity, which is best understood in terms of the fact that they are invariant in a number of respects. In the remainder of this section I will examine these invariances in some detail.

The idea of spelling out the modal aspects of laws of nature in terms of invariance is not new; Mitchell (2003, 140), Lange (2009) and Woodward, to name a few, have done so before. I will not discuss any of these approaches in any detail. Let me, however, mention Woodward (1992; 2018), whose view is probably closest to the one presented here. He explicitly endorses the idea that nomological necessity should be understood as an invariance claim: 'We may say that a law, in contrast to an accidentally true generalization, expresses a relationship which not only holds in the actual circumstances but which will remain stable or invariant under some fairly wide range of changes or interventions' (Woodward 1992, 202).

Invariance is clearly a modal notion. It concerns not only actual but also counterfactual changes. While it might be argued that moving from one modal notion (nomological necessity) to another (invariance) is not much progress, there is certainly at least one advantage. The notion of invariance naturally leads to a closer examination of the modal structure delineated by the internal generalisations. Invariance is a *relative* notion, and we thus have to ask, 'Invariance with respect to what?' Furthermore, as we will see later, the concept of invariance helps to understand how modal notions can be empirically accessible.

While Woodward has briefly alluded to some distinctions among invariance relations (2003, chapter 6; 2018), there has been no systematic examination.

We can specify the following invariance relations, which will shed light on the inviolability of laws or their nomological necessity:

1) *Invariance of the law equation with respect to initial conditions.* The law equation holds irrespective of which values from a certain range of initial conditions, or boundary conditions or other sets of variables, obtain or would obtain. Newton's second law, a dynamic law, was supposed to hold for any place and any initial velocity characterising a system. By contrast, the ideal gas law holds only for a restricted range of values of p and V (when read as determining T). Both law equations are invariant with respect to at least some initial conditions; note that this is an invariance with respect to the *values* of variables that explicitly figure in the law equation.

The other two kinds of invariances that I will discuss concern invariances with respect to features of the world that are not represented as variables in the law equation.

2) *Invariance of the law equation with respect to other features of the systems.* When we determine the speed of a free-falling body there are a number of properties that can be ignored, such as the shape and the colour of the falling body. That is what Galileo's law implies by not mentioning them. The fact that certain variables do not figure explicitly in the law equation (and cannot be determined by those explicitly mentioned) implies that the law equation is invariant with respect to these variables. The law equation in the ideal gas law is in practice understood as saying not only that the temperature of actual ideal gases is determined by the pressure and the volume but also that there is no other feature that might be relevant: neither smell, nor the shape of the molecules, and so on.

   In the case of macroscopic laws, two kinds of properties with respect to which invariance may occur ought to be distinguished:
   a. Same-level properties; for example, colour, shape, mass in the case of free-falling bodies.
   b. Lower-level or constitutional properties; for example, the molecular structure of the gases in the case of the ideal gas law. This kind of invariance plays a major role in discussions about universality in the context of phase transitions. Two different kinds of

micro-physical invariances are relevant. First, macroscopic physical properties may be invariant with respect to changes of the system's dynamical state on the micro-level. Second, the macrobehaviour might even be invariant with respect to non-actual counterfactual changes in a system's composition at the microlevel. For instance, in a ferromagnetic system one might add nextnearest neighbour interactions to a system originally having only nearest neighbour interactions and scale down the strength of the original interaction in a manner that would leave the macroscopic magnetization of a system invariant (see Hüttemann, Kühn and Terzidis 2015).

Whether or not a law equation is invariant with respect to features of the system not represented in the law equation is a matter that needs to be established empirically. If such invariances obtain, we are justified in abstracting away from the relevant features of the system in the law equation we use to describe the system's behaviour. It may, of course, happen that as a result of empirical investigations, certain invariance claims with respect to other features of the system under consideration have to be given up. That will lead to law equations that contain more variables. Thus, for instance, the ideal gas equation was at some point replaced by the van der Waals equation, the Peng–Robinson equation, and so on, equations that take into account some features of the molecules that constitute the gas.

3) *Invariance of the law equation with respect to the behaviour of other systems in the universe.* Newton's second law adequately describes the behaviour of any (physical) body irrespective of the behaviour of other systems in the universe. This is not to say that other systems cannot have an influence on the system we attempt to characterise in terms of Newton's second law but rather that the law equation continues to hold for the body despite external forces being impressed on it. Whatever the forces are that affect a certain system, Newton's second law will continue to hold for the system under investigation.

The status or role of the three kinds of invariance relations is somewhat different. The first kind of invariance (invariance with respect to initial conditions) merely expresses the fact that the law equation holds for more than one set of values of the relevant variables. This is part of the content of what the law explicitly says about the behaviour of systems. The second kind of invariance (invariance with respect to characteristics of the system that are not represented in the law equation) provides a rationale for abstracting in the law equation from the features in question. The third kind of invariance (invariance with respect to the behaviour of other systems in the universe) seems to be most relevant for accounting for the inviolability or natural necessity of laws. If this kind of invariance holds, then whatever changes there are, the law equations will remain the same.

There is an interesting and important further contrast between the invariance of the law equation with respect to other features *of the same system* and the invariance of the law equation with respect to the behaviour of *other systems in the universe* when it comes to evidence for the failure of invariance. Whereas in the former case, as already indicated, the law equation will be revised (in the light of sufficient evidence), in the latter case, one strategy is to hedge the law by a ceteris paribus clause. Instead of revising the law equation in Galileo's law when it comes to falling objects in water or other media, it may be argued that the law holds ceteris paribus. I will deal with this issue in Chapter 2.

Let me add a few remarks about the notion of invariance.

First, I agree with Woodward that invariance-based accounts of nomological necessity 'provide a naturalistic, scientifically respectable and non-mysterious treatment of what non-violability and physical necessity amount to' (Woodward 2018, 160). Invariance claims can be scientifically investigated. For example, if it is argued that a certain law equation is invariant with respect to the colour of the system, it is reasonably clear what is claimed and we know how to check the claim. Furthermore, macroscopic invariance claims (e.g., the invariance of the behaviour of gases with respect to their constitution) can at least in principle be explained in terms of lower-level laws (as well as experimentally investigated).

Let me add that I see no reason to believe that there is a special problem with knowing invariance facts simply because they are modal facts. Here is, for instance, how Galileo's spokesman Salviati, on the basis of experiments, argued for the claim that bodies with different densities would fall with equal speed in a vacuum, that is, that their speed is invariant with respect to the density of the bodies:

> We have already seen that the difference of speed between bodies of different specific gravities is most marked in those media which are the most resistant: thus, in a medium of quicksilver, gold not merely sinks more rapidly than lead but it is the only substance that will descend at all; all other metals and stones rise to the surface and float. On the other hand, the variation of speed in air between balls of gold, lead, copper, porphyry, and other heavy materials is so slight that in a fall of 100 cubits a ball of gold

would surely not outstrip one of copper by as much as four fingers. Having observed this I came to the conclusion that in a medium totally devoid of resistance all bodies would fall with the same speed. (Galileo 1954, 71–2)

Salviati's inference becomes problematic only if one already starts out with the idea that our epistemic access is limited to the actual. Laws of nature make modal claims that are empirically accessible.

Second, this latter claim is further illustrated by the fact that additionally, some more specific notions of invariance play an explicit role in physics. Some laws or theories are characterised as Galilei invariant, Lorentz invariant or Gauge invariant. These concepts point to symmetries in the systems under investigation, symmetries with respect to certain classes of transformation. Not every law or theory is Lorentz invariant, and whether or not the behaviour of systems can be characterised as Lorentz invariant is an empirical matter. The notion of invariance that is relevant in these physical discussions is the same as I used previously; that is, the law equations continue to hold under certain kinds of actual and counterfactual changes. More specifically, the homogeneity of space, for instance, is a symmetry in the sense that the laws of classical physics remain the same under translations in space; all space points are equivalent when it comes to Newton's second law or the Schrödinger equation. Absolute space points turn out to be irrelevant for the dynamics of systems (Castellani 2003, 429). The same holds for Lorentz transformation, and so on. Laws are invariant if they stay the same under actual or counterfactual changes. Some of these invariances, namely those listed here, are constitutive of what is usually considered to be nomological necessity. Others, such as Lorentz invariance, are additional invariances that laws may or may not comply with.

A third remark addresses an objection against analysing lawhood or nomological necessity in terms of invariance relations. Psillos criticises Woodward's account for being circular because the notion of invariance presupposes that of a law. Lawhood, Psillos argues, should thus not be explicated in terms of invariance. Rather, 'some laws must be in place before, based on considerations of invariance, it is established that some generalization is invariant under some intervention' (Psillos 2002, 185). The disagreement concerns the question of whether lawhood or invariance should be taken to be the fundamental notion. I have argued that we have good reasons why we should explicate lawhood or nomological necessity in terms of invariance (rather than the other way around), because we thus understand the role nomological necessity plays in scientific practice.

In dealing with the circularity objection, it is important to distinguish an ontological issue and an epistemological issue. The ontological issue concerns the question of what the invariance of a law equation with respect to, say, the behaviour of other systems *consists in*. The simple answer is that the equation continues to hold when there are actual or counterfactual changes in the behaviour of other systems. There is no need to refer to laws of nature when it comes to explicating what invariance consists in.

The epistemological issue is a different issue. It concerns the question of how we might know that a certain invariance claim is true: how we come to know that a certain law equation would continue to hold given changes in the behaviour of other systems, or – to use Psillos's phrase – how the invariance claim is 'established'. In order to decide whether or not a certain generalisation is invariant with respect to certain changes we may indeed rely on laws. But that is no problem for the invariance account as long as the ontological and epistemological issues are kept apart. Thus, laws may play a role in establishing invariance claims, but that does not imply that the ontological characterisation of invariance presupposes the notion of a law.[15]

1.3.4 *External Generalisations and Modal Surface Structure*

Let me now turn briefly to external generalisations and the question of whether their role in scientific practice requires additional modal assumptions. In Section 1.1.3 I argued that the explanation of why the stone has a certain velocity presupposes that the system in question, the free-falling stone, falls under Galileo's law. The truth of the external generalisation, I argued, is a presupposition for the internal generalisations doing their explanatory work. Furthermore, in the case of our example of standard explanation, all that is required is that the external generalisation is true. Thus, standard explanation, which is of course only one aspect of scientific practice, does not commit us to postulating modal structure, let alone *additional* modal structure.

The case is different when it comes to prediction or manipulation. Let us go back to our preliminary law statement:

(A) 'All systems of a certain kind K behave according to Σ.'

---

[15] Woodward mixes these two issues up by defining invariance in terms of interventions (see, e.g., Woodward 2003, chapter 6).

As already indicated, in predicting, manipulating or constructing systems we rely on laws, and this reliance can be explicated in terms of the invariance of the law equation with respect to actual or counterfactual changes in the behaviour of other systems in the universe. However, the fact that the law equation continues to characterise the behaviour of the system under consideration presupposes that the external generalisation (A) is invariant with respect to the same changes. No matter what other systems are doing, all systems of kind K will continue to behave according to Σ. So, when it comes to manipulation or prediction, external general-isations commit us not only to their truth but also to their invariance with respect to the behaviour of other systems in the universe. There is, however, no evidence that we are committed to any new kind of invariance relations that we have not encountered in our discussion of internal generalisations.

Let me close this section on external generalisations by giving a diagnosis as to why the modal surface structure that is delineated by law statements has often been ignored. Two facts seem to me to be relevant here. First, the reconstruction of law statements along the lines of the 'All ravens are black' paradigm has encouraged us to overlook the role of internal generalisa-tions. Second, when scientific practice was analysed at all, it has often been assumed that description and explanation is all there is. As I have argued, we can understand the role of external generalisations with respect to description and explanation without making modal assumptions, since the truth of the external generalisation suffices in these cases. It is only when we consider other scientific practices, such as predicting or manipu-lating, or the role internal generalisations play in these practices, that we are forced to consider the modal surface structure of law statements.

1.4   Conclusion

The purpose of Section 1.3 to was to examine the modal surface structure of laws, that is, those modal assumptions that best account for how we make use of external and internal generalisations in scientific practice. I argued that in order to understand not just our explanatory practice but also other practices that involve a reliance on laws, such as predicting, manipulating or constructing systems, we have to make two assumptions. First, we have to assume that law statements attribute a space of possible states to systems. Second, we must assume that both the law equations and the external generalisations are invariant with respect to certain actual and counterfac-tual changes. The law equation is invariant with respect to (i) a range of initial conditions and (ii) features of the system that do not figure in the law equation. In addition, both the law equation and the external generalisation are invariant with respect to changes of the behaviour of other systems in the universe. These empirically accessible invariance claims account for what is usually taken to be the law's nomological necessity.

As already mentioned, it is important that invariance claims are modal claims. It is claimed not only that law equations are invariant with respect to actual changes but also that they are invariant with respect to counterfactual changes. Some may feel that for this reason we have made little progress. Nomological necessity, they will argue, is mysterious precisely because it is a modal notion; invariance is a modal notion too, so we are left with a mystery. I think this is wrong. Being able to characterise the modal structure that comes with law statements in more detail and explaining how invariance claims are empirically accessible does seem to me to constitute progress.

Finally, as mentioned before, I am exclusively concerned with the modal surface structure of laws, which the notion of invariance helps to characterise. One may then still ask 'What underpins these invariances?' (Bird 2007, 5). That may be an interesting question, but it is one that transcends the approach taken in this book. As I said at the beginning, I want to confine myself to metaphysical claims that can be established via an inference to the best explanation of why we have the scientific practice we have. Assuming that laws are invariant in various respects does exactly that. Invariance relations, as I will show in the following chapters, can account for (almost) all the natural modalities we encounter in scientific practice. A further analysis of invariance in terms of subjunctive facts (Lange), essences (Bird), dispositional modality (Mumford) or the Humean mosaic does not do any extra work in explaining scientific practice. A minimal metaphysics of scientific practice should thus abstain from such hypotheses of how natural modality is to be further explained or reduced. Those who do conduct such investigations have to appeal to intuitions about the governing of laws, intuitions about the supervenience or non-supervenience of laws on the underlying Humean mosaic, or intuitions about quiddities, or to a preference for desert landscapes. It is hard to see an argument as to why such intuitions or preferences should be considered to be truth conducive. Thus, I am quite happy with the modal surface structure for which I have argued. 'Those who go beneath the surface do so at their peril' (Oscar Wilde).

CHAPTER 2

The Problem of Ceteris Paribus Clauses

Scientific practitioners are proper parts of the world and typically explain, predict or manipulate the behaviour of systems, which are proper parts of the world, too. Scientific practice takes place *within the universe*. That is to say, in order to understand this practice, it is not sufficient to consider a *View from Nowhere* or the view of a mathematical archangel who might use generalisations for the sole purpose of registering and systematising what has happened, is happening and will happen in the universe as a whole. Rather, as I argued in the previous chapter, the inward-looking scientist's perspective needs to be replaced by the perspective of scientists within our world to understand the scientific practices we have. Practitioners use generalisations *within the universe* for specific purposes, which might concern both actual and non-actual possible states of the universe. More particularly, generalisations allow us to extrapolate what we have come to know in one situation to other situations, e.g., in prediction or manipulation. The analysis of the practice of extrapolation will reveal further metaphysical assumptions underlying the use of law statements.

In predicting and manipulating the behaviour of parts of the universe we treat these systems as if they were the whole world and abstract away from the rest of the universe (cf. van Fraassen 1989, 218; Scheibe 1991b, 343–4). It is essential for understanding scientific practice to answer the question of why we can successfully do science by disregarding most of the universe. All the evidence we have for our best theories and laws is from observation and manipulation of parts of the universe. In fact, it may be conceptually impossible to do experiments on the universe as a whole (see Breuer 1995 for discussion). If the only proper system of investigation were the universe as a whole, we could probably not have any evidence for laws or theories that pertain to it. Thus, even if it were to turn out that there is a sense in which the universe is the only 'proper' system theories or law statements refer to, there needs to be an account of (a) why our investigation of parts of the universe works as evidence for such laws and (b) why law statements are sometimes useful for manipulating or explaining parts of the universe. The universe as a whole may be an ideal system in the sense that there are no external factors that interfere with the behaviour that a law statement might attribute to it, but we also need to know how to understand the scientific practice that takes part in the world. It is thus essential for our understanding of laws that they can be used for explaining, predicting or manipulating the behaviour of subsystems of the universe.

Subsystems qua being *sub*systems are prone to being interfered with by other subsystems. This interference can take two different forms. In the first case, other subsystems may have an influence on *the values of the variables* that figure in a law equation but not on the law equation itself. For instance, systems in the environment may exert forces on the system we are interested in. However, whatever the forces are, the law equation, e.g., in Newton's second law, will continue to hold for the system under investigation. The law equation remains invariant with respect to actual or counterfactual changes in other systems.

In the second case, other subsystems not only influence the values of the variables that figure in the law equations but furthermore, interfere with the system such that the law equation no longer holds. The law equation in Galileo's law no longer adequately characterises the behaviour of a falling body if a medium, e.g., water, interferes with a falling object. In this chapter, what I am interested in is the second sort of situation. In such cases, the law equation seems to be no longer invariant with respect to changes in the behaviour of other systems – even though in Chapter 1 we considered this kind of invariance as being constitutive for nomological necessity and thus for lawhood.

2.1 Ceteris Paribus Laws and the Problem of Extrapolation

Not all laws that play a role in the explanatory practice in the sciences are invariant with respect to changes in the behaviour of other systems of the universe. Galileo's law for falling bodies describes the behaviour of falling bodies, but only provided they fall in a vacuum. The Schrödinger equation with the Coulomb potential describes the behaviour of hydrogen atoms, but only provided no external fields are present. Newton's first law ('N1') describes the behaviour of bodies, but only provided no forces are applied on them:

> (N1) Every body continues in its state of rest or of uniform motion in a straight line, unless it is compelled to change that state by forces impressed upon it. (Newton 1999, 416)

Laws that describe the behaviour of systems *provided certain conditions obtain* are usually called ceteris paribus laws (henceforth: cp-laws). These laws were originally introduced in (medieval) economics (see Reutlinger, Schurz, Hüttemann and Jaag 2019). The role of these laws in biology, psychology, the social sciences, etc. has been extensively discussed in the last decades (e.g., Fodor 1974; Popper 1974; Johansson 1980; Kim 1985; Lepore and Loewer 1987; Fodor 1991; Beatty 1995; Carrier 1998; Mitchell 2002; Kincaid 2004 and Roberts 2004).

Laws that hold only provided certain conditions obtain, i.e., cp-laws, exist not only in biology or economics but also in physics. In what follows I want to argue that the objections usually raised against treating cp-laws as respectable items in the natural sciences can be overcome. An analysis of extrapolation will be particularly helpful to uncover certain methodologies and their presuppositions, which at the same time will serve to explicate the role of exclusive cp-laws. I will also point to implications this has for other disciplines.

I will focus on non-lazy, exclusive cp-laws. Exclusive cp-laws state that systems display a certain behaviour *provided there are no disturbing factors*, whereas comparative cp-laws require that certain (often unspecified) factors remain *constant* (see Schurz 2002 for this distinction). A cp-clause is *lazy* if all exceptions to the law (or disturbing factors) can be listed and it is merely a matter of convenience and the result of 'laziness' that the conditions are not listed explicitly (see Earman, Roberts and Smith 2002, 283 for this distinction).

In the next two sections I will outline the main problems that accounts of cp-laws have to face. With respect to one of these problems, viz. providing truth conditions for cp-laws, I defend a dispositional account and argue that traditional objections can be met. This will involve a somewhat lengthy digression into what I call the 'practice of extrapolation' (Sections 2.4 and 2.5). Next, I explain how the other major problem can be solved, i.e., how cp-laws can be tested (2.6). After dealing with some cases that proved difficult for other accounts of cp-laws (2.7), I will briefly return to the question of why in scientific practice we can successfully disregard most of the universe and focus on parts of the universe. In Section 2.8, I discuss how to understand the modal aspects of dispositions in the light of the metaphysics developed so far. Finally, in Section 2.9, I briefly compare my account of laws with other dispositional accounts of laws, most notably Bird's (2007).

2.2 Provisos, Dilemmas and Other Frustrations

What is wrong with exclusive cp-laws? There are different diagnoses. Some authors argue that there are no cp-laws to start with, while others point to severe difficulties and dilemmas for cp-laws. In this section, I will argue for two claims: (1) there are cp-laws (in a certain sense); (2) cp-laws are confronted with at least two major problems.

The problems concerning cp-laws are usually introduced by way of a dilemma. Many laws, such as Galileo's law, are false if the law is read as a strict (universal) generalisation, which is invariant under changes in the behaviour of other systems. The claim 'Whenever a body falls, it falls according to the equation $s = \frac{1}{2} gt^2$' is false, because in water and other media the equation does not correctly describe the behaviour of the bodies in question. Similarly, the claim 'Hydrogen atoms obey the Schrödinger equation with a Coulomb potential' is false if read as a strict generalisation, because there may be electrical or magnetic fields. That is the first horn of the dilemma. If, on the other hand, the law is hedged by a *ceteris paribus* clause (henceforth cp-clause), Galileo's law becomes 'Whenever a body falls, it falls according to the equation: $s = \frac{1}{2} gt^2$, *unless some interfering factor intervenes.*' (Restricting the law to *free*-falling bodies is hedging, too.) Similarly, the claim 'Hydrogen atoms obey the Schrödinger equation with a Coulomb potential, *unless some interfering factor intervenes*' has been hedged. These claims appear to be trivially true, at least as long as the notion of an interfering factor is not specified further. If what is meant by an interfering factor is simply 'a factor that makes the law turn out to be false', the hedged claim says no more than 'The relation $s = \frac{1}{2} gt^2$ holds, unless it does not.' Making similar observations, Marc Lange concludes: 'For many a claim that we commonly accept as a law statement, either that claim states a relation that does not obtain, and so is false, or it is shorthand for some claim that states no relation at all, and so is empty' (Lange 1993, 235).

Without the cp-clause the laws in question are false; if, on the other hand, a cp-clause ('provided there are no interfering factors') is added, the laws become empty or trivially true. (In what follows, I will call this the 'Lange's dilemma'.)[1] Most philosophers wish to avoid this dilemma. The attempt to do so reveals the need for an account of the truth conditions of

---

[1] Cartwright discusses a slightly different dilemma: 'Ceteris paribus generalisations, read literally without the "ceteris paribus" modifier, are false. [...]. On the other hand, with the modifier the ceteris paribus generalisations may be true, but they cover only those few cases where the conditions

cp-laws: an adequate account of such laws ought to explicate that these law statements can be true and empirically testable statements.

But is this a plausible way of setting up the problem? One might doubt whether the correct way is to ask whether there really is a dilemma for cp-laws. Earman and Roberts argue to the contrary. Dealing with their objections to Lange's dilemma will help to differentiate different kinds of problems and dilemmas that cp-laws are confronted with.

Earman and Roberts do not deny that cp-*clauses*[2] play a role in scientific practice. What they deny is that these clauses should be taken to be part of the laws. Rather, these clauses come into play when the laws are *applied* to particular systems. Thus, the application of the law is hedged by a cp-clause, not the law itself. (Another way of putting the same point: changes in the behaviour of other systems may affect the values of the variables that figure in a law equation, but never the law equation itself.)

Earman and Roberts's argument can be illustrated by one of the examples I introduced earlier. I took the claim 'Hydrogen atoms behave according to the Schrödinger equation with the Coulomb potential' (henceforth '(H)') to be a cp-law because it is true only provided there are no further fields, etc. According to Earman and Roberts the law in question is not (H) but the bare Schrödinger equation, which is strictly true without any qualification. In scientific practice, the Schrödinger equation is applied to a particular kind of system (hydrogen atoms). Equations for the behaviour of hydrogen atoms can be derived, provided one is able to specify the Hamiltonian. It is in the process of specifying the Hamiltonian for the system under consideration that one has to assume that there are no further forces or fields besides the ones mentioned in the Hamiltonian. Earman and Roberts endorse the claim made by Hempel that when theories or laws are applied to particular systems one has to presuppose a proviso to the effect that the system is subject to no additional forces. But this proviso or cp-clause is not part of the law, i.e., the Schrödinger equation. The latter is strictly true and not qualified by a proviso. Furthermore, according to Earman and Roberts, the claim 'Hydrogen atoms behave according to the Schrödinger equation with the Coulomb potential' is not even a law. It is a statement about a particular kind of system that is derived from the law (Schrödinger equation) and some additional premises. Among these additional premises is the claim

are right' (Cartwright 1983, 45). In this case, the horns of the dilemma are falsity and restricted applicability rather than falsity and triviality.

[2] In what follows, I will use 'cp-clause' and 'proviso' (a term used by Hempel (1988)) interchangeably.

that the system can be completely characterised in terms of a particular Hamiltonian. This premise may or may not be true, but its truth does not affect the truth of the law, i.e., the Schrödinger equation. The content of the law should not be confused with the equations that are derived from the law and some additional assumptions. Since (H) is not a law, it is therefore not a cp-law. Similarly, Newton's first law is not a law and thus not a cp-law. It can be derived from a law, namely, Newton's second law, with the additional assumption that the net force on the system under investigation is zero.

So Earman and Roberts's proposal is to restrict the honorific title 'law' to Einstein's gravitational field law, to the Schrödinger equation or to Newton's second law. These are meant to be strictly true, i.e., without cp-clauses. There is thus no Lange's dilemma for these fundamental laws. Since all the other generalisations (Newton's first law, Galileo's law and special science generalisations) are only erroneously given the title 'law', they cannot be cp-*laws*, and thus Lange's dilemma does not apply to them, either.

Before I point out where I disagree with Earman and Roberts's approach, I need to make clear that I am not committed to the view that *all* laws or generalisations in physics are cp-laws. I agree with Earman and Roberts that

> typical theories from fundamental physics are such that *if*[...] they were true, there would be precise proviso-free laws. For example, Einstein's gravitational field law asserts – without equivocation, qualification, proviso, ceteris paribus clause – that the Ricci curvature tensor of spacetime is proportional to the total stress-energy tensor for matter-energy. (Earman and Roberts 1999, 446)

However, having said this, there remains one essential point of disagreement with Earman and Roberts's assessment: It is still a problem that scientists use generalisations that hold only provided interfering factors are absent – not only in biology and economics but also in physics. (H) is an example. The problem is this: Without a cp-clause (H) is false; if, on the other hand, a cp-clause ('provided there are no interfering factors') is added, (H) appears to become empty or trivially true. Lange's dilemma remains.[3] Depriving (H) of the honorific title 'law' does not provide a solution to Lange's dilemma. Whether or not we decide to call these generalisations '(cp-)laws', we need to understand how they can successfully play the role they play in prediction, explanation or manipulation. It might furthermore be argued that Lange's dilemma remains a problem for the bare Schrödinger equation as well: All the empirical evidence we have for it comes from applications that are described by

---

[3] The fact that (H) is false in the presence of interfering factors does not depend on whether the influence of these interfering factors is small or large.

generalisations such as (H). Thus, we cannot understand how the Schrödinger equation can be tested, i.e., we cannot make sense of an essential aspect of scientific practice, unless we understand how generalisations such as (H) work.

The problem of explaining how to make sense of the cp-clauses is thus not a side issue. Every account of laws has to make sense of the fact that law statements describe the behaviour of systems that are embedded in larger contexts, which potentially interfere with the law equation in question.

The aim of the following sections is to show that Earman and Roberts's pessimistic assessment concerning the debate of cp-law statements, namely, that

> [. . .] there is no persuasive analysis of the truth conditions of such laws; nor is there any persuasive account of how they are saved from vacuity; and, most distressing of all, there is no persuasive account of how they meld with standard scientific methodology, how, for example, they can be confirmed or disconfirmed. In sum, a royal mess. (Earman, Roberts 1999, 470–1)

can be rejected.

An account of such 'cp-laws', as I will continue to call them, has to address at least two problems:

First, the *Semantic Problem*: What are truth conditions for cp-laws? Even if we were to decide to no longer call generalisations like (H) 'laws', we would have to be able to specify their truth conditions. Without the cp-clause the generalisations in question are false; if, on the other hand, a cp-clause ('provided there are no interfering factors') is added, the generalisations appear to become empty or trivially true (Lange's dilemma). Any account of cp-laws has to overcome this dilemma. The first step is to spell out the truth conditions for cp-laws.

Second, the *Confirmation Problem*: How can these generalisations or cp-laws be confirmed or disconfirmed? Here the problem is that, as long as the notion of an interfering factor is not constrained, such generalisations appear to be immune to disconfirmation. Corresponding to Lange's dilemma, there is a confirmation dilemma: Without the cp-clause the generalisations in question will almost certainly be disconfirmed; if, on the other hand, a cp-clause ('provided there are no interfering factors') is added, the generalisations appear to become immune to disconfirmation.

2.3 Pietroski and Rey's Proposal

From Section 2.4 onwards I will develop and defend a dispositional account of cp-laws. I will argue (1) that such an account can deal with the above-mentioned problems and (2) that certain objections that have been raised against dispositional accounts can be countered.

Before developing this account, however, I will briefly discuss the view of Pietroski and Rey, because it contains some ideas I would like to build on.

Pietroski and Rey (1995) are concerned with what I have called the semantic problem. They deny that cp-laws are necessarily trivial. For the cp-law to be non-trivial in spite of counterinstances we need independent evidence for the existence of an interfering factor. It would not be acceptable if the only evidence for the existence of the disturbing factor were the counterinstance to the law we started with. Their essential requirement is the following:

(P&R) For all x, if Ax, then (either Bx or there exists an independently confirmable factor that explains why ¬Bx)

The important point is that the notion of an interfering factor has to be made more precise without simply requiring a list of possible interfering factors. The second disjunct in condition (P&R) (the case of ¬Bx) does not cite potential disturbing factors explicitly; rather, it involves second-order quantification over potential interfering factors (Pietroski and Rey 1995, 93). The second-order quantification makes the notion of an interfering factor more precise in the case of non-lazy cp-laws.

Unfortunately, there are well-known counterexamples to Pietroski and Rey's account.[4] Woodward has pointed out that the generalisation 'All charged objects accelerate at 10 m/s^2' qualifies as an acceptable cp-law according to Pietroski and Rey's account. Let us assume that there are cases such that charged objects do indeed accelerate at 10 m/s^2. Furthermore, if a charged object fails to accelerate at 10 m/s^2, this can always be explained in terms of an independently confirmable factor such as the presence of an electromagnetic field (Woodward 2002, 310). However, the recipe works for *any* value of the acceleration. As a consequence, we have a plethora of cp-laws 'All charged objects accelerate at *n* m/s^2' for any *n*. This seems to be a counter-intuitive result.

Earman and Roberts (1999, 453) discuss the counterexample 'If something is spherical it is electrically conductive.' We can explain the fact that some round objects fail to be electrically conductive in terms of an independently confirmable factor, namely, the molecular structure of the objects in question.

---

[4] See Reutlinger (2011) for a partial defence of Pietroski and Rey.

The problem that becomes apparent in both cases is the following: Pietroski and Rey's conditions do not guarantee that, given the absence of interfering factors, the antecedent is nomologically sufficient for the consequent. Being spherical does not have anything to do with electrical conductivity. And the fact that an object has a certain electrical charge may be relevant for acceleration but not sufficient for the occurrence of one particular value of acceleration. The properties referred to in these laws are only accidentally related.

2.4 The Semantics of CP-Laws

My proposal for the semantics of cp-laws will draw on suggestions that have already been discussed in the literature (Mill 1836; Cartwright 1989; Hüttemann 1998; Lipton 1999; Bird 2005; Hüttemann 2014). These suggestions, however, need to be augmented in order to be plausible vis-à-vis certain objections that have been raised.

One of these authors (John Stuart Mill) explicitly denies that laws have exceptions:

> There are not a *law* and an *exception* to that law – the law acting in ninety-nine cases, and the exception in one. There are two laws, each possibly acting in the whole hundred cases, and bringing about a common effect by their *conjunct operation* [my italics, A.H.]. [. . .] Thus if it were stated to be a law of nature, that all heavy bodies fall to the ground, it would probably be said that the resistance of the atmosphere, which prevents a balloon from falling, constitutes the balloon as an exception to that pretended law of nature. But the real law is that all heavy bodies *tend* to fall [. . .]. (Mill 1836, p. 56, original emphasis)

Reading 'laws with exceptions' as 'exclusive cp-laws', Mill's suggestion can be reformulated as a pair of claims:

(a) Exclusive cp-laws ought to be read as strict laws about tendencies.
(b) If there appears to be an exception, i.e., if the tendency is not manifest, this should be understood as an effect of a 'conjunct operation' of more than one tendency.

In the remainder of this (rather long) section, I will argue for (a) and (b) (in a rather long digression about the practice of extrapolation). More particularly, I will argue that this practice is best explained by assuming that laws attribute multi-track dispositions (or tendencies) to systems. Assuming (a) and (b) is the best way to make sense of this aspect of scientific reasoning. At the same time, the analysis of extrapolation allows us to make Mill's notion of a 'conjunct operation' sufficiently precise that a standard objection against dispositionalist accounts of cp-laws can be met. Once these metaphysical assumptions are established, we can solve both the Semantic Problem and the Confirmation Problem.

2.4.1 Dispositions

As I already indicated, my account of the role of cp-laws in scientific practice will comprise claims about dispositions. So, I should say a few things about what I take dispositions to be. My general strategy is to commit myself only to features of dispositions that are explanatorily relevant for understanding scientific practice. Only three features of dispositions will play a role in this section: Dispositions are properties, they can be distinguished from categorical properties and they may be multi-track. I will deal with modal aspects of dispositions in Section 2.8.

Let me first turn to the question of how dispositional properties can be distinguished from categorical properties. Courage, malleability and solubility are usually considered to be dispositional properties. By contrast, squareness and other geometrical properties are often considered to be the paradigmatic candidates for categorical properties. It has been found to be notoriously difficult to draw a distinction between dispositional and categorical properties (see Mumford 1998, 64–92 or Choi and Fara 2016). In this chapter I will try to get by with assumptions about dispositional and categorical properties that are fairly uncontroversial.

A useful starting point is Matthew Tugby's characterisation of dispositionalism and dispositions:

> According to dispositionalism, dispositions (or what are sometimes called 'causal powers') are taken to be real properties of concrete things, and properties which cannot be reduced to any more basic kind of entity. But what, precisely, does it mean to say that a property is irreducibly dispositional in nature? Roughly, it means that the property is characterised in terms of the causal behaviour which things instantiating that property are apt to display. This is to say, in other words, that irreducibly dispositional properties are by their very nature orientated towards certain causal manifestations. To illustrate: in order to explain what it means for, say, a particle to have the property of being charged, the dispositionalists will point out for example that charged particles accelerate when placed in an electro-static field. In explaining this feature of charge, dispositionalists take themselves to have said something about the essential nature of charge. In short, then, dispositionalists see properties as properties for something else: their causal manifestations. (Tugby 2013, 452)

This is a useful starting point because it allows me to start with a disagreement: There is no need to consider the relation between a disposition and its manifestation as *causal*. Although 'causal' appears in Tugby's abstract characterisation of dispositions, the example (charge) does not rely on causal terminology: An object has the property of being charged if – given the right conditions – it accelerates in the right way. In what follows I will work with a notion of disposition that does not build the notion of causation into it. Assumptions about causation are additional assumptions and are not necessary for a characterisation of dispositions (There is a further reason for this approach: As I will argue in the next chapter, causation can be explicated in terms of dispositions. For this to be non-circular, dispositions had better be explicable without recourse to causal terminology.)

As a consequence, I start with a slightly modified version of Tugby's rough characterisation: A dispositional property is characterised in terms of the behaviour which things instantiating that property are apt to display. They are 'oriented towards their manifestation', as Tugby notes, which implies that the manifestation need not be displayed. Dispositional properties thus allow a distinction to be drawn between a property being *instantiated* and a property being *manifest*. A person, for instance, can be brave without actually displaying brave behaviour. Non-dispositional or categorical properties, by contrast, do not allow this distinction: a piece of paper cannot instantiate the property of squareness without displaying squareness. In this case, it seems inappropriate to consider such a distinction. If we furthermore assume that the distinction between categorical and dispositional properties is exhaustive, we can derive a simple and uncontroversial sufficient condition for the ascription of dispositional properties: If the ascription of a property assumes that the property in question can be instantiated without the manifestation being displayed, it is a dispositional property. Whatever the exact nature of dispositions may be, this simple criterion will allow us to identify some dispositions. Thus, we have a sufficient condition for identifying dispositional properties.

A further feature of dispositional properties that I will appeal to is that they might be multi-track. I take multi-track dispositions to be dispositions that cannot be characterised in terms of one stimulus condition only or one manifestation only (for a similar characterisation see Vetter (2015, 40)). Courage, for instance, is best characterised as a multi-track disposition. Courage may manifest in different ways depending on the context the person instantiating the property is in. Still, it is usually conceived of as a single disposition. Multi-track dispositions will be relevant in what follows in at least two contexts. For instance, the different values of variables that are used to describe different initial conditions of the temporal behaviour of a system can be thought of as characterising different tracks. The same applies to different kinds of interferences. An interesting question – one I will return to later – is what justification we have for the assumption that these are tracks of one and the same disposition: Why is this one multi-track disposition rather than a bundle of different dispositions?

Malleability is another example of a multi-track disposition. A piece of gold is malleable because it may take various shapes depending on how it has been, e.g., hammered. The case of these two multi-track dispositions (courage and malleability) differ in an interesting way that is relevant for the preceding criterion for identifying dispositions. Courage is unproblematic with respect to our criterion. A person may be brave without actually displaying any brave behaviour. The situation is, however, different in the case of malleability. Every piece of gold will certainly have some shape or other. It seems that necessarily exactly one manifestation of its malleability is displayed. Does this provide a counterexample to our criterion (because it is not true that malleability may or may not be manifest)? That is not the case if we understand the criterion as follows: a property is dispositional if the property can be instantiated even though for every single track it is true that its manifestation need not be displayed. Even if the particular shape, i.e., the particular manifestation of the disposition, had not been manifest, the piece of gold would have had exactly the same multi-track disposition (malleability) that it actually has. So, even though a piece of gold necessarily has some shape in virtue of its malleability, it need not have been this particular shape. Just like other dispositional properties, malleability allows a distinction between a property being instantiated and a property being manifest, and that means in our context a distinction between a multi-track disposition being instantiated and its being manifest in accordance with one particular track.

In the next few sections, I will argue that the problem of providing a semantics for cp-laws can be solved by assuming the existence of dispositional properties. Before I present the argument, I would like to make a few remarks about what I do *not* commit myself to in relying on dispositions. A major line of reasoning concerning dispositions in the metaphysics of science starts with assumptions about the identity of properties. Shoemaker argued that 'the identity of a property is completely determined by its potential for contributing to the causal powers of the things that have it' (Shoemaker 1980, 133). This line of reasoning has been taken up by, e.g., Ellis (2001) and Bird (2007), identifying causal powers with dispositions (Chakravartty 2007, 123). The main claim is that 'properties have essences that are dispositional in nature' (Bird 2007, 43). I will defend no particular view about the identity of properties. In positing dispositions, I am thus not committed to any form of essentialism or to the claim that dispositions should be understood as causal powers (as already mentioned). Let me furthermore add, though this is disconnected with the previous line of reasoning, that I am not committed to the view that a disposition needs a stimulus condition to become manifest; the absence of interfering factors might suffice for manifestation.

The only features I will appeal to in this section are the following: Dispositions are properties, they may be multi-track, and they can be distinguished from categorical properties because they allow a distinction to be drawn between a property being instantiated and a property being manifest.

2.4.2   The Problem of Extrapolation I

Let me now turn to extrapolation. As I already mentioned, the analysis of this practice is central to the metaphysical claims in this chapter. Once we have identified the metaphysical assumptions that best explain this practice, we can deal with both the Semantic Problem and the Confirmation Problem for cp-laws.

So, what is extrapolation? In a book on evidence-based policy, Nancy Cartwright and Jeremy Hardie discuss the following question. A project carried out in Tennessee in the 1980s provided empirical evidence that students in smaller classes did better than students in larger classes and that this applied in particular to minority students. In the mid-1990s California intended to apply a class size reduction programme to its state schools. Would what had worked in Tennessee work in California too? This question concerns the issue of whether evidence gained in one case is relevant for a different case in another context (Cartwright and Hardie 2012, 4).

The problem Cartwright and Hardie are concerned with is finding evidence for whether or not it is rational to infer from one case to the other. That is an epistemic problem. It is not the problem I am interested in. There is, however, a further issue. Assume that the inference works. Why does it work? The 'why' in this question is not asking for the evidence we have but rather for the metaphysical assumptions that best account for the success of extrapolation. The problem of extrapolation, as I will discuss it, consists in the challenge of explaining why, in a case of successful extrapolation, generalisations that have been found to hold under specific circumstances also hold under different circumstances. (See Steel (2008, 3) for examples and a slightly different characterisation of the problem.) The problem of extrapolation thus concerns the metaphysical assumptions that best explain the success of a certain aspect of scientific practice.

The (metaphysical) problem of extrapolation is not only relevant for social policy or in medical research but also in very simple examples from physics, to which I will turn now. Take Galileo's law again. It describes the behaviour of free-falling bodies, i.e., the behaviour of a body falling in a vacuum. But what about falling bodies in a medium? As a matter of fact, we successfully consider the vacuum case to be relevant for the other cases as well: We take it as the basis of our theoretical treatment of the other cases and add further influences (further terms). So, in a successful extrapolation, what we know about one kind of case is relevant for other cases with different contexts. What accounts for this successful practice? Extrapolation from, in some sense, ideal situations to less ideal situations is just one specific example of the practice of extrapolation. The general question to answer is what metaphysical assumptions best explain why it is possible to extrapolate from one case to another. In virtue of what can we extrapolate from one subsystem of the universe in a particular context to another subsystem in a different context?

My main metaphysical claim in what follows (on which the solution to both the Semantic Problem and the Confirmation Problem for cp-laws relies) is that the success of the practice of extrapolation is best explained by assuming that law statements attribute special kinds of dispositional properties. In a first step, I will show why extrapolation presupposes dispositional properties. In a second step, I will characterise these dispositions in more detail.

Cartwright in another publication (where she was concerned with the metaphysics rather than with the epistemology of extrapolation) suggested the following explanation of what is going on:

> When [...] disturbances are absent the factor manifests its power explicitly in its behaviour. When nothing else is going on, you can see what tendencies a factor has by looking at what it does. This tells you something about what will happen in very different, mixed circumstances – but only if you assume that the factor has a fixed capacity that it carries with it from situation to situation. (Cartwright 1989, 191)

In my terminology, Cartwright argues that we need dispositions to understand why extrapolation works. When we consider Galileo's law of free fall ('nothing else is going on, disturbances are absent') to be relevant for a falling body in a medium (different context, 'mixed circumstances'), we assume that something carries over from the first (ideal) situation to the second situation: a property that is manifest in the first situation but fails to be manifest in the second (though, of course, it is instantiated in both situations).[5]

Though this argument maybe read as a transcendental argument, it need not be reconstructed that way.[6] The claim that dispositional properties are required for an explanation of the success of the practice of extrapolation can be broken down into two separate assumptions. The first is that something 'carries over' from the one situation to the other. This assumption is best read as being backed by an inference to the best explanation. There are two qualitatively different situations: situation A (no disturbances) and situation B (mixed circumstances). As a matter of fact, we take it that situation A tells us something about situation B (that is the uncontroversial explanandum). A straightforward explanation for this fact is that there is some respect in which the two situations agree. It is an identical feature of the systems in question by virtue of which we can extrapolate from one situation to the other. In other words: We are dealing with the *same property* in both situations, and that accounts for one aspect of the practice of extrapolation, viz. the carrying over.

It seems that the most natural reading of 'same property' in this context is to conceive of it as a universal, that is, as something that is wholly present and thus numerically the same in both situations. Assuming that the systems in the different contexts instantiate the same universal explains why situation A tells us something about situation B, i.e., why we can move from one context to another.

---

[5] There is a tradition – starting with Cartwright's own considerations on the reality of component forces – according to which – contrary to what I argue – the property in question is manifest in both situations. Thus, Cartwright writes: 'the manifestation [. . .] is fixed even if the behaviour described in occurent-property language [. . .] is highly varied' (Cartwright 2008, 195). However, that does not solve our original problem; it reappears under a different guise. We now have to account for the fact that there is a manifest property in different situations that gives rise to different displays of occurent behaviour. It seems that we need (given we agree that the property in question is manifest) a manifest property that may or may not display a certain occurent behaviour. (For a detailed discussion of a 'triadic' conception of dispositions, see Fischer 2018, 84–99.)

[6] Anjan Chakravartty (2017, 119–20) argues that when it comes to extrapolation an analysis in terms of dispositions is not *required* – meaning it is not the only possible explanation of this practice. While true, this is an objection against a transcendental reading of the argument only but no objection if the argument is reconstructed as involving an inference to the best explanation.

Is this the only conceivable understanding of 'same property'? Maybe not. One might argue that extrapolation is due to an inference based not on the identity of an underlying universal but on the (perfect) similarity of some feature of A and some feature of B. These features may be thought of as non-universal particularised features of systems (tropes). I don't think that the features of scientific practice that we have looked at commit us to one of these understandings of what it is for systems to have the same property.

What is essential for understanding how extrapolation works is that the carrying over from one situation to the other is best explained by the fact that the systems in both situations share the same property – whatever the underlying metaphysical nature of this property may be.

The second metaphysical assumption involved in the claim that in order to make sense of extrapolation we have to assume dispositional properties is the claim that the properties in question are dispositional rather than categorical properties. As I explained earlier, if a property allows being instantiated but fails to be manifest it qualifies as a dispositional property. This is indeed the case when it comes to extrapolation. The behaviour that is manifest in the vacuum situation is not manifest if a medium is present. If the property that carries over from one to the other situation, i.e., the property that is instantiated in both situations is manifest in one of these but fails to be manifest in the other, it cannot be a categorical property; it has to be a dispositional property. So, this second metaphysical assumption is not backed up by an argument (whether transcendental or inference to the best explanation;) it simply follows from the application of the sufficient condition for identifying dispositional properties.[7]

2.4.3 The Problem of Extrapolation II

In what follows I will characterise the dispositions in question in more detail by looking at what happens in situations in which interfering factors

---

[7] According to Steel (2008, 5), postulating disposition or capacities is not merely not the only conceivable explanation for the problem of extrapolation; it is not even the best explanation. Note, however, that his 'problem of extrapolation' is slightly different from the one discussed here. Steel is interested in what we best rely on in situations in which we try to inductively infer in what is going on in heterogeneous situations. Merely knowing that a capacity or disposition is present in the different situations is of no great help (see also the criticism by Earman and Roberts discussed in Section 2.4.3). More detailed knowledge is necessary, for instance by marrying the dispositions approach to mechanisms (Steel 2008, 79). By contrast, I am interested in the explanation of what we should assume about the nature of the properties involved given the extrapolation strategy was successful. The argument is that these properties should be conceived of as dispositional properties. That is compatible with the claim that this information is of no great help when we are faced with a concrete problem of making an inference with respect to heterogeneous situations.

are present. The manifestation of the behaviour of the falling body and the medium together provide an instance of a 'conjunct operation', to use Mill's terminology. The behaviour to which the body is disposed is not manifest due to the presence of the medium – however, the body and the medium *contribute* to the behaviour of the compound and are thus what might be called 'partially manifest'.

So, the claim is that if we take the law statements to be attributing dispositions, we understand why the law statement can be extrapolated, e.g., to systems in non-ideal situations, that is, to situations in which further factors are present. Something (a dispositional property) is present in both situations. The fact that we are dealing with the same property in both situations accounts for the fact that what we know in the one situation is relevant for the other situation as well. The fact that the property is dispositional has to be assumed because in (at least) one of these cases the property isn't (completely) manifest.

But how *exactly* does the ascription of dispositions help to explain why we may extrapolate, for example, from, say, an ideal case to a less ideal case? The worry is that understanding laws as disposition ascriptions might not explain the *actual* behaviour of a system. Earman and Roberts, for example, ask:

> Thus if what one wants explained is the actual pattern, how does citing a tendency – which for all we know may or may not be dominant and, thus, by itself may or may not produce something like the actually observed pattern – serve to explain this pattern? (Earman and Roberts 1999, 451 f).

The fact that something (the disposition) carries over does not tell us enough as long as it isn't clear *how* the disposition contributes to the phenomenon in question/the behaviour that arises in new contexts in which (further) interfering factors are present.

If we just had Galileo's law, it would indeed be difficult to see how the relevant disposition would help to explain what is going on in non-vacuum situations. However, within Newtonian mechanics (as well as in quantum mechanics) we have the means to describe precisely how different dispositions contribute to the actual patterns of observed behaviour in the presence of other systems such as interfering factors. Within Newtonian mechanics we can describe the vacuum case as a system consisting of the earth's gravitational field and the body within this field. Here again, it might be argued that knowledge of how this system (and the body as part of it) would behave if it were on its own does not tell us anything about how it actually behaves in the presence of other factors, for example a medium. For all we know, it might be argued, the other factors might or might not be dominant. But things are not as gloomy as this objection implies. We know how to determine what is happening in such cases. We have the means to determine how the various factors contribute. Whether or not one tendency is dominant is not an accidental or inexplicable fact.

There are at least two ways to treat the fall in a medium, both of which allow us to specify how a disposition contributes to actual patterns of observed behaviour. According to the first and more rigorous treatment, we consider a compound system consisting, e.g., of the falling object and all the molecules that make up the medium as well as all their interactions plus the external gravitational field. Thus, according to the first account, we are relying on part-whole explanations in order to determine precisely how different dispositions contribute.

Example: Let me present one example of how part-whole explanation accounts for how dispositions contribute. The fact that quantum mechanics has a law of combination or composition is not very often made explicit. An exception is the physicist Arno Bohm, who mentions as one of the basic assumptions or axioms of quantum mechanics:

> IVa. Let one physical system be described by an algebra of operators, A_1, in the space R_1, and the other physical system by an algebra A_2 in R_2. The direct-product space $R_1 \otimes R_2$ is then the space of physical states of the physical combinations of these two systems, and its observables are operators in the direct-product space. The particular observables of the first system alone are given by $A_1 \otimes I$, and the observables of the second system alone are given by $I \otimes A_2$ (I = identity operator). (Bohm 1986, 147)

This is quantum mechanics' law of composition ('COMP' for further references) for the case of non-identical physical systems (or non-identical 'particles'). It is only in virtue of this law that part-whole explanations in quantum mechanics are possible. (The metaphysical status of COMP will be discussed after the presentation of the example.)

I will illustrate this explanatory strategy through a simple example that does not take interactions into account. The example in question is the explanation of the energy spectrum of carbon monoxide (CO): carbon monoxide molecules consist of two atoms of mass m_1 and m_2 at a distance x. Besides vibrations along the x-axis, they can perform rotations in three-dimensional space around its centre of mass. This provides the motivation for describing the molecule as a rotating oscillator rather than as a simple harmonic oscillator. The compound's (the molecule's) behaviour is explained in terms of the behaviour of two subsystems, the oscillator and the rotator. (The two subsystems are not spatial parts but rather sets of degrees of freedom.) Bohm, who discusses this example in his textbook on quantum mechanics, describes this procedure as follows:

> We shall therefore first study the rigid-rotator model by itself. This will provide us with a description of the CO states that are characterised by the quantum number n=0, and will also approximately describe each set of states with a given vibrational quantum number n. Then we shall see how these two models [the harmonic oscillator had already been discussed in a previous chapter of Bohm's book] are combined to form the vibrating rotator or the rotating vibrator. (Bohm 1986, 128)

Thus, the first step consists in considering how each subsystem behaves if considered as an isolated system. The second step consists in combining the two systems by relying on COMP.

Bohm considers the following subsystems: (i) a rotator (the rotational degrees of freedom of the CO molecules), which can be described by the Schrödinger equation with the Hamiltonian: $H_{rot} = L^2/2I$, where L is the angular momentum operator and I the moment of inertia; (ii) an oscillator (the vibrational degrees of freedom of the CO molecules), which can be described by the Schrödinger equation with the following Hamiltonian: $H_{osc} = P^2/2\mu + \mu\omega^2 Q^2/2$, where P is the momentum operator, Q the position operator, ω the frequency of the oscillating entity and μ the reduced mass.

He adds up the contributions of the subsystem by invoking COMP. Thus, four laws are involved in this part-whole explanation:

(1)   The law for the compound (the CO molecules): The compound behaves according to the Schrödinger equation with the Hamiltonian $H_{comp} = H_{rot} + H_{osc}$. (This is the explanandum)

(2)   The law for the rotator (the rotational degrees of freedom of the CO molecules): The rotator behaves according to the Schrödinger equation with the Hamiltonian: $H_{rot} = L^2/2I$.

(3)   The law for the oscillator (the vibrational degrees of freedom of the CO-molecules): The oscillator behaves according to the Schrödinger equation with the Hamiltonian: $H_{osc} = P^2/2\mu + \mu\omega^2 Q^2/2$.

(4)   The law of composition (COMP) that tells us how to combine (2) and (3).

We explain the behaviour of the compound as described in (1) – the explanandum – in terms of (2), (3) and (4) – the explanans.

How does this help with Earman and Robert's problem? (2) and (3) – according to our interpretation – should be read as attributing dispositions to the rotator and oscillator, respectively. In a compound system (the CO molecules), neither the behaviour referred to in (2) (I am thinking here in particular of the energy spectrum that follows from these attributions) nor the behaviour referred to in (3) is manifest. But the appeal to (2) and (3) nevertheless explains the behaviour of the compound (in particular the energy spectrum) via (4), i.e., via the law of composition. Appeal to (2), (3) and (4) explains in what sense (1) is a 'conjunct operation' in Mill's sense.

The case considered is particularly simple because the parts do not interact. This is generally not the case. For instance, if we were to treat the falling body in terms of a part-whole explanation, we would have to specify every single molecule constituting the medium as well as all their interactions plus the external gravitational field.

The second, simplified treatment of 'conjunct operation' in general and the case of a falling stone in a medium adds an additional force term to the net force that acts on the falling object. This extra force term represents the buoyancy, which is the effective force resulting from all the molecules that constitute the medium. (This second simplified treatment does take for granted the mathematical structure that allows for part-whole explanations.) Thus, in this simple case, the two force terms represent what the falling body and the medium are disposed to. The vector addition law then determines how what the systems are disposed to translates into actual behaviour.

Both these treatments give a quantitative account of how the disposition of a system, e.g., the falling body – as manifest in the vacuum case – partially accounts for (and thus contributes to) the actually observed pattern or behaviour in the presence of an interfering factor such as the medium. The essential point is that there are laws or rules that allow us to quantitatively determine how different factors or tendencies contribute to the actual behaviour that we want to explain. Whether or not one tendency 'dominates' is thus perfectly explicable on a dispositional account. In other disciplines, such as, say, economics, there might be no general rules for dealing with disturbing factors, but there might be more local rules that allow us to represent interfering factors, e.g., how an economy is affected by decisions of the federal reserve bank.

What we have seen so far is that understanding laws as attributions of dispositions does not undermine their ability to explain actual behaviour – provided there are laws of composition. But what is the metaphysical status of laws of composition? These meta-laws had better not be understood as attributing dispositional properties. That would give rise to a regress because it would not be clear how they contribute unless there are meta-meta-laws. We thus should assume that the laws of composition cannot be interfered with, that instantiation guarantees manifestation, i.e., that they attribute categorical properties or behaviour. This assumption, however, does not seem to be problematic because there is no evidence that they can be interfered with. The laws of composition are invariant with respect to changes in the behaviour of other systems.

Even if this is settled, there are two slightly different interpretations of the relation of the laws of composition and the dispositions involved.

The first interpretation of laws of composition and (a fortiori, as we will see) of dispositional properties follows Mill and Cartwright (though she does not believe in general laws of composition – see Cartwright 1983) and has been advocated in some of my earlier writings. There is a tendency, capacity or disposition that is fully revealed in ideal situations, i.e., in situations in which interfering factors are absent. In the presence of interfering factors, i.e., in the presence of other dispositions, the original disposition *contributes* to the overt behaviour; it is not completely manifest but only *partially*. Complete manifestation is manifestation *simpliciter*. While 'contribution' and 'partial manifestation' may sound metaphorical at first, what is meant by these terms has been made precise by the laws of composition, as we have seen. There are two features that are distinctive for this first interpretation. The ideal situation, i.e., the absence of interfering factors, has a special status; it is here that the disposition is fully revealed, i.e., completely manifest. The second feature is the status of the laws of composition. They have the status of being special meta-laws that determine how various dispositions exercise their 'conjunct operation'.

A second interpretation of laws of composition and, a fortiori, of dispositions takes the ideal and the non-ideal cases, i.e., the cases of complete and partial manifestation, to be simply different tracks of a multi-track disposition. To recapitulate: A multi-track disposition is a disposition that cannot be characterised in terms of one stimulus condition only or one manifestation only. The fragility of a porcelain vase will be displayed (slightly) differently depending on whether it is hit by a hammer or whether it is crushed by a car driving over it. Still, we believe that fragility is one disposition rather than a bundle of separate dispositions – in this case presumably by virtue of its underlying molecular structure, which we make responsible for both 'tracks'.

The assumption that laws attribute multi-track dispositions can be motivated as follows: When we characterise systems as classical mechanical or quantum mechanical systems it seems artificial to give the laws of composition a separate status. The law of composition quoted earlier is an integral part of quantum mechanics. In light of this consideration, we ought to read law statements as attributing the whole structure of Newtonian or quantum mechanics to the systems in question. When we claim that hydrogen atoms behave according to the Schrödinger equation with the Coulomb potential we assume that all of quantum mechanics fully pertains to hydrogen atoms. The laws of composition are part and parcel of this attribution.

This has consequences for how to understand the disposition that is attributed by a law statement. Let me go into a little more detail here. By virtue of allowing for a large range of initial conditions, as discussed in Chapter 1, we have already accepted that law statements attribute many 'tracks'. Taking into account that law statements attribute dispositional properties, we have to conclude – already on the basis of the fact that laws attribute spaces of possible states – that law statements attribute multi-track dispositions. For example, a law statement like 'Ideal gases behave according to the equation $pV=\nu RT$' implies an infinite number of 'tracks' for every single system:

If $p=p_0$ and $V=V_0$, T will take the value T_0
If $p=p_1$ and $V=V_1$, T will take the value T_1
If $p=p_2$ and $V=V_2$, T will take the value T_2
...

These tracks are generated within Σ because the internal generalisations range over different values of the variables. In particular, there is a different track for every value of the pressure p. Similarly, in the case of Galileo's law we have different conditionals (or tracks) for different values of t. In the case of dynamical equations, typically different initial conditions give rise to different tracks of the attributed disposition.[8]

On the present interpretation of the laws of composition, besides the domain internal generalisations quantify over, there is another mechanism that generates tracks of a disposition. If we read law statements as attributing all of, e.g., quantum mechanics to the system in question when it is asserted that hydrogen atoms behave according to the Schrödinger

---

[8] The term 'multi-track disposition' is used by some authors as requiring different *types* of manifestation – whereas I require different manifestations (or different trigger conditions) only (see, e.g., Kistler 2012). As a consequence, on my account – but not on Kistler's – determinable properties come out as multi-track properties. Nothing hinges on these terminological choices.

equation with the Coulomb potential, we see that a second mechanism for generating tracks is due to the laws of composition. The behaviour of the atoms is conditional on the presence of interfering factors or additional subsystems:

Track 1:   If no interfering factor is present, the hydrogen atom will indeed behave according to the Schrödinger equation with the Coulomb potential,

Track 2:   In the presence of interfering factors of kind I, it will contribute to the overall behaviour of the compound system depending on the exact constitution of the factor and the interactions.

etc.

So, according to this interpretation, law statements attribute multi-track dispositions to systems. Some of these tracks obtain in virtue of the fact that the variables on which the behaviour depends can take different values. There is, however, another dimension of tracks because the contribution will differ depending on the presence of interfering factors. It is this latter dimension of tracks that is relevant for the practice of extrapolation. Extrapolating from the ideal case to a non-ideal case is – according to the interpretation of composition laws and dispositions under consideration – an extrapolation from one track to another track of a multi-track disposition.

Even though it is useful to distinguish these two mechanisms of track generation, it should be noted that this distinction is not dictated by nature. Whether or not an additional force, say on an electron, is treated as an interfering factor (mechanism of the second kind) or simply as a change in an initial condition (mechanism of the first kind) depends on how we choose to represent the situation, i.e., how we carve out the system we are interested in. If our focus is on the electron itself, the force may be represented as an external interfering factor. We might, however, be interested in a more inclusive system, say the electron in its environment, and treat the force as a special initial condition of this system.

In the light of what we have said about extrapolation and the role of laws of composition, three issues might be raised with respect to the dispositions under consideration: (1) Why should we consider the different tracks to be tracks of one multi-track disposition rather than a bundle of different dispositions? (2) Do these multi-track dispositions comply with our original characterisation of dispositions in Section 2.4.1? (3) Is the disposition finitely characterisable?

First: Why multi-track dispositions? The alternative is to assume a bundle of distinct dispositions (and thus a bundle of distinct laws). Some authors firmly embrace the claim that what we are dealing with is a bundle of independent laws. Armstrong, for instance, holds:

> We have to say, I think, that functional laws are bundles of what we might call *particularized laws,* laws that hold for determinate properties, determinate forces, masses, and distances in the case of the gravitation law. The determinables, very important properties but not universals, when suitably connected by some mathematical relationship, give, as it were, 'instructions' for the particularized laws where the work of the world is done. (Armstrong 2010, 43)

According to the bundle view, every particular regularity implied, e.g., by the ideal gas equation is a law of its own, but properly speaking there is no gas law (at the level of determinables). Given our assumption that law statements attribute dispositions, Armstrong's view implies that there is a bundle of distinct dispositions rather than a multi-track disposition when it comes to functional laws.

It could be argued that the same holds when we are dealing with interfering factors. According to the bundle view, we are dealing with a plethora of particularised regularities:

(i) $\forall x \, (Cx \, \& \, \text{no other interfering factor} \supset M_0x)$
(ii) $\forall x \, (Cx \, \& \, I_1x \, \& \, \text{no other interfering factor} \supset M_1x)$
(iii) $\forall x \, (Cx \, \& \, I_2x \, \& \, \text{no other interfering factor} \supset M_2x)$
(iv) $\forall x \, (Cx \, \& \, I_3x \, \& \, \text{no other interfering factor} \supset M_3x)$
. . .
(. . .) $\forall x \, (Cx \, \& \, I_1x \, \& \, I_2x \, \text{no other interfering factor} \supset M_{12}x)$
. . .

Here, Cx stands for system x being in circumstances C, I_ix for x being affected by an interfering factor and M_ix for x displaying behaviour M.

Every particularised regularity should then be read as attributing a distinct disposition to system x. Bird (2007, 22) proposes this view.

Such a reading, however, gives rise to what might be called the problem of discreteness (see Schrenk 2010 for raising the 'problem of discreteness' in a slightly different context). These laws and their underlying dispositions are unconnected: Why should (i) be explanatorily relevant for (iv)? Why should evidence for (i) be evidence for (iv), or vice versa? These questions are particularly pertinent when we re-examine the carrying-over argument, which we discussed in Section 2.4.2. The argument simply wouldn't work with the bundle view. We considered two different situations, which we called *situation A* (no disturbances) and *situation B* (mixed circumstances). Because different behaviours are displayed in the two different situations, the bundle view will argue that two different dispositions are present in situation *A* and situation *B*. But then we are confronted with the question of why we consider what we learned about situation *A* (about one kind of disposition) to be relevant for situation *B* (where we are allegedly confronted with a different kind of disposition). Taking scientific practice as an indicator of the underlying structure of reality requires a unity of the different tracks. The carrying-over argument requires multi-track dispositions and does not work with the bundle view. (Armstrong, by contrast, argues for the bundle view by relying on metaphysical principles and intuitions only; see Armstrong 2010, 21–2; 42–3).[9]

The second issue arises for both these interpretations of laws of composition. Suppose we say of a hydrogen atom that it is disposed to behave according to the Schrödinger equation with the Coulomb potential. This behaviour will be either completely or partially manifest. So, it will be manifest (in some sense) under all circumstances. Put differently, necessarily, one of the above-mentioned regularities ((i), (ii), etc.) or tracks will be manifest. Shouldn't we then say that according to our classification this is a categorical property?

There is an easy answer if we give a special role to the ideal track. We can then say that what is relevant is complete manifestation. It is in virtue of the fact that the behaviour in question can fail to be completely manifest that the criterion for dispositionality is met: the attributed property can be instantiated without being manifest (that is, completely manifest). This reply makes use of the fact that complete manifestation is distinguished, because in the ideal situation, as opposed to all the others, the disposition is undisturbed – there are no interfering factors. According to the other interpretations, we just have multiple tracks, none of which has a special status. So, this reply is not available. We are considering a situation in which a multi-track disposition is such that necessarily one of the many manifestations is displayed. So, there is a sense in which the disposition is necessarily manifest. Should we then classify it as a categorical property? No, rather – as in the case of malleability discussed earlier – we should say that the property that is attributed by a law statement allows for

[9] My argument for multi-track dispositions (as I understand them) implies that determinable properties should be considered to be irreducible to bundles of determinate properties – against, e.g., Armstrong's view quoted in the preceding passage (see Wilson 2012 for discussion).

a distinction between the property being instantiated and the property being manifest, and that means in our context that for every track of the disposition, if the particular track of the disposition had not been manifest, the system under consideration would still have had exactly the same multi-track disposition it actually has. The property in question thus allows for a distinction between the property being instantiated and the property being manifest in this specific manner. That is to say, it is a dispositional property because it can be instantiated without the manifestation for track t being displayed, and the same holds for every other manifestation.

The third issue is whether multiple-track dispositions can be finitely characterised if there are infinitely many tracks. That, however, is only a problem if we assume that every track has to be listed explicitly. We have already seen that this need not be so, for example in Pietroski and Rey's case of a second-order quantification (Section 2.3). We have also seen in the case of the ideal gas laws that mathematical variables allow us to characterise infinitely many situations without explicitly listing every single one of them. In the case of the multi-track dispositions that we are discussing here, infinitely many tracks (i.e., infinitely many different combinations of kinds of interfering factors) can be characterised in terms of the laws of composition because they provide the means to deal with any interfering factor or subsystem.

What both interpretations of the laws of composition and thus of the dispositions share is that there is one disposition – one single disposition – which, according to the first interpretation, is completely or partially manifest, or which, according to the second interpretation, is a multi-track disposition. The unity of the disposition has to be assumed to make sense of the practice of extrapolation, which requires the assumption that something (i.e., one thing) is present in various situations.

I do not think much depends on whether we choose the first or the second interpretation. The second interpretation seems to me to be – metaphysically speaking – preferable because it does not give a special status to laws of composition and does not mark ideal situations as being ontologically distinguished (which is, of course, compatible with them being of special epistemological significance). However, the second interpretation has the disadvantage of making the presentation of what follows rather cumbersome. In what follows in the remainder of this chapter I will thus stick for most of the time with the first interpretation and hope that nothing hinges on this choice.

Let me summarise: The success of extrapolating from one situation to a qualitatively different situation is best explained in terms of a property of the system in question that is present in both situations and that thus accounts for the extrapolation. As we have seen, this property is dispositional: If the situations are qualitatively different (in the sense that the observed behaviour that is to be explained is different), then in at least one of the cases the property is not (completely) manifest. A fortiori, the properties that account for extrapolation are dispositional properties according to the criterion discussed in Section 2.4.1.

In these cases, according to the foregoing argument, our standard characterisation of law statements

(A) All physical systems of a certain kind behave according to Σ.

should be read as (or replaced by):

(B) All physical systems of a certain kind *are disposed to behave* according to Σ.

Let me make two additional remarks:

(1) The preceding reconstruction of law statements as attributing dispositions gives rise to a third kind of counterfactual, which is neither an *other object counterfactual* nor a *same object counterfactual* of the sort introduced in Section 1.1.2 but rather, a counterfactual typically associated with interferences into the display of a disposition: If the system were isolated it would behave according to Σ. If the system were interfered with by a subsystem of kind K, the compound system would behave according to Σ', where Σ' is a different specialisation of the theory in question. (These kinds of counterfactual will be essential for understanding causal reasoning; see Chapter 3.)

(2) It should be noted that the argument from extrapolation does not show that *all* law statements attribute dispositional properties to systems – but those that describe a behaviour that can be interfered with do. There are some examples of law statements for which we have no reason to assume that the attributed behaviour can be interfered with. The claim that all systems obey the bare Schrödinger equation is such an example. While the claim certainly does not characterise systems completely, it does constrain the behaviour of systems. The law statement seems to be true under all circumstances; the systems cannot be interfered with such that the law equation does not hold.

2.4.4   The Solution of the Semantic Problem for cp-Laws

We have seen that many law statements should be read as attributing dispositional properties to systems. That offers a solution for some of the problems raised for exclusive cp-laws.

If cp-laws (in the sense of law statements) can be taken to attribute dispositions to physical systems, we can easily provide a semantics for cp-law(-statements) so as to avoid the Lange dilemma. According to this account, a cp-law(-statement) is true provided the type of system in question has the disposition that the cp-law statement attributes to it.

cp: All As behave according to Σ (where 'Σ' stands, e.g., for the Schrödinger equation with a particular Hamiltonian)

is explicated by

All As are disposed to behave according to Σ.

Reconstructing law statements as statements about dispositions rather than about overt behaviour turns cp-laws into strict laws, because all As without exception have the disposition. In other words: If the law statement is reconstructed in terms of a disposition ascription, the law equation is invariant with respect to changes in the behaviour of other systems. Lange's dilemma does not apply. The cp-clause is needed at the level of overt behaviour; its function can be explicated in terms of dispositions underlying the overt behaviour where no such clause is necessary.

2.5   An Objection to the Dispositional Account and a Reply

The dispositional account thus provides a neat solution to Lange's dilemma. The dilemma does not even arise, because what the law says ('All As are disposed to behave according to Σ') is no longer hedged by a cp-clause.

However, there is an obvious and well-known problem with this account. The problem concerns the question of whether the dispositional account can indeed avoid the dilemma for exclusive cp-laws. Even though it is true that the laws on this construal no longer appeal *explicitly* to cp-clauses, Lange's dilemma seems to turn up in a different place (Lipton 1999, 166). The dispositionalist has to specify the triggering conditions for the disposition in order to give the disposition ascription a determinate content. The conditions under which the disposition will manifest itself are exactly those that require explicating the cp-clause (Lipton 1999, 166–7;

Schrenk 2007a). The dispositionalist reconstructs the law 'cp, all As behave according to Σ' as: 'All As are disposed to behave according to Σ.' This is informative only to the extent that the triggering conditions for the disposition can be spelled out. So here we are again – or so it seems. Something informative must be said about the cp-clauses. By way of response I will distinguish two ways to spell out the manifestation conditions for the dispositions in question (which I have not distinguished up to this point). According to the first option, *the disposition to behave according to Σ* becomes manifest provided there are no *interfering* factors. According to the second option, *the disposition to behave according to Σ* becomes manifest provided there are no *other* factors at all (whether interfering or not), i.e., if the system in question is *on its own* – if it constitutes its own world, so to speak. The first option is problematic because something informative must be said about interfering factors and must be integrated into the specification of the disposition. As long as the notion of an interfering factor can be spelled out as 'whatever prevents the manifestation of the disposition', the disposition ascription is analytically true and thus trivial. The second option is more promising, as the notion of an *interfering factor* is not an integral part of the specification of the disposition. Rather, the disposition becomes manifest if the system in question is on its own. The problematic notion of an inferring factor can thus be avoided.

One might object that this move is of no great help because we still have to deal with an indefinite number of factors that need to be absent. Rejoinder: Talking about '*interfering* factors' generated Lange's dilemma. The problem was that if what is meant by an interfering factor is simply 'a factor that makes the law turn out to be false', a cp-claim would say no more than 'Every A is B, unless it is not' (see Section 2.2). By being able to avoid the term 'interfering factor', we now have a precise statement of under what conditions the systems in question will show the attributed behaviour. Still, there is a kernel of truth in the objection: According to the second reading, the dispositions become manifest in certain *ideal* situations that are never realised (see also Mumford 1998, 87–91 for a discussion of ideal circumstances in the context of disposition ascriptions). But how can these ideal situations be relevant for the real world given that there is an indefinite number of factors, which undermine the system being on its own? The answer is that the ideal situations make explicit how the system would behave if it *were* on its own. In the discussion of the extrapolation problem (Section 2.4.1), we have seen that the ideal case is relevant for the non-ideal case because the ideal behaviour *contributes* to the real situation and that laws of composition give us a precise account of this relevance.[10]

One last question remains: How can the law statements that attribute these dispositions be *tested* in the presence of (a possibly indefinite number of) other factors? This is an important epistemic problem but does not concern the *semantics* of the cp-laws. I will now turn to this problem, the Confirmation Problem. (Solving the Confirmation Problem does not directly contribute to the metaphysics of laws. So, the reader might want to go directly to Section 2.8. However, the preceding account of laws in terms of dispositions would not be plausible if it remained unclear how law statements understood as ascriptions of dispositions can be tested in scientific practice – and how, in the light of such an account, traditional counterexamples to other accounts can be dealt with.)

2.6   Testing cp-Laws

A dispositional account of cp-laws is a tenable option only if it can explain how cp-laws can be tested. This is what I earlier called the 'Confirmation Problem' for cp-laws. In what follows, I will present an account of the epistemic acceptability of cp-laws from a dispositionalist perspective (which is not meant to imply that other accounts of cp-laws cannot accommodate the procedures I present).

We have already seen that Pietroski and Rey's account (read as an account of epistemic acceptability) has been confronted with various counterexamples. In this section I will argue that the confirmation problem can be solved by augmenting this account. More specifically, I will argue that epistemically acceptable cp-laws need to be embedded in an experimental and/or theoretical methodology that allows us to determine the influence of the interfering factors. Not only must the independently confirmable factor explain why the consequent does not occur; it must furthermore be possible to vary it experimentally or theoretically so that we can determine how the system would behave in the absence of disturbing factors.

---

[10] This reply to the objection depends on the first reading of disposition, which ties its (complete) manifestation to ideal situations. On the second, that is, the multi-track reading, all tracks (whether ideal or less than ideal) are on a par, so this reply is not available. However, on the second reading, the laws of composition are part and parcel of what is attributed by the disposition-attributing law statement. Therefore, there is no need to explicitly list the absence or presence of all possible interfering factors as triggering conditions. The laws of composition are able to comprise an infinity of tracks and thus an infinity of possible interfering factors or additional subsystems without explicitly listing any of them.

2.6.1 Example

Let me illustrate my claim by discussing two candidates for cp-law statements:

(N1) Every body continues in its state of rest or of uniform motion in a straight line, unless it is compelled to change that state by forces impressed upon it. (Newton 1999, 416)

According to the confirmation dilemma, we are confronted with the following problem: Without the phrase 'unless it is compelled to change that state by forces impressed upon it' it can easily be disconfirmed; with the phrase it appears to be immune to disconfirmation, because it seems to say not much more than 'unless it does not'. Every counterinstance may be immunised by postulating the occurrence of this or that force.

If this assessment were correct, we would have exactly the same evidence for the following completely fictitious alternative to Newton's first law:

(N1*) Every body rotates, unless it is compelled to change that state by forces impressed upon it.

Both N1 and N1* are hedged; they contain explicit cp-clauses. Both are furthermore on a par in that it is unlikely that we will ever come across a manifestation of the respective disposition.

Nevertheless, we think of Newton's first law as true,[11] whereas we hold N1* to be false. Intuitively we think that N1 and N1* are not on a par with respect to the evidence we have for them. Thus, the assessment of the epistemic situation mentioned earlier must be wrong.

So, what went wrong? The short answer is that we have not taken into account that there are methodologies that allow us to test apparently immunised claims such as N1 or N1*. We accept N1 rather than N1* because N1 can be confirmed and N1* can be disconfirmed. Physics provides methodologies for testing cp-laws. The methodology allows us to determine what would happen if the interfering factors were absent. In every case of an interfering factor, the influence on the system's behaviour can be made quantitatively precise. This allows us to see whether the nomological connections postulated in N1 and N1* are true or false.

The quantitative determination of the influence of interfering factors can proceed by experimental or theoretical means. To give an example: A charged object may fail to manifest the behaviour attributed by N1 due to the presence of a magnetic field. If we are able to manipulate the field, we

---

[11] For the purposes of this chapter, I disregard the fact that Newton's mechanics has been replaced.

will see that the behaviour of the charged object approximates more and more the behaviour postulated in N1 if the force on the electron is minimised. This confirms N1. By contrast, minimisation of external forces on the charged object yields no approximation to the behaviour postulated in N1*. This disconfirms N1*.

An experimental manipulation of a disturbing factor will not always be possible. However, there are also theoretical means to determine the influence of disturbing factors. If, for instance, a comet moves in the gravitational field of the sun, Newton's second law plus the law of gravitation will tell us how the comet would move if the mass of the sun were minimised and how it would move if there were no mass at all. On the basis of these laws we are able to determine that the comet would move according to N1 rather than according to N1*.[12]

The rationale behind both the theoretical and the experimental procedure is that what happens in the ideal circumstances, where the disturbing influences do not operate, continues to be a contribution to what happens when some disturbing influences are also operating. In moving from the ideal to the real world we add in the disturbances using the relevant law of composition.

2.6.2 *Epistemic Acceptability*

On the basis of the above considerations, I suggest that we should complement Pietroski and Rey's conditions for epistemic acceptability of cp-laws. Pietroski and Rey required (among other things):

(P&R) For all x, if Ax, then (either Bx or there exists an independently confirmable factor that explains why ¬Bx)

As we have seen, this condition is not sufficient for epistemic acceptability. It needs to be supplemented by the following condition:

– Based on experimental or theoretical methods there is evidence for how the behaviour of the system under consideration is affected by the disturbing factor and that in the limit of a disappearing influence of the disturbing factor A is nomologically sufficient for B, i.e. $(A \rightarrow B)$ is a law.[13]

Such evidence presupposes that the quantitative influence of the disturbing factor in question can be determined, such that it is known how the system

---

[12] In this particular case, the result may be trivial because assuming the framework of Newton's theory ensures that N1 comes out as confirmed – but that is not true in general.

[13] For a similar suggestion, see Schrenk (2007b, 158–61).

would behave in the absence of the disturbing factor. The important point is that for a cp-law to be epistemically acceptable, the potential disturbing factors need not be cited explicitly – it is not required that a finite list of factors can be presented.

I have developed this augmented condition for the epistemic acceptability from the perspective of the dispositional account of cp-laws. The condition, as well as the methodologies described that serve to illustrate the condition, provides evidence for the claim that not only the semantic but also the confirmation problem can be solved by a dispositional approach.

2.7 Dealing with Counterexamples

In this section, I will show that the augmented conditions for the epistemic acceptability of cp-laws enable us to deal with the counterexamples by Woodward as well as by Earman and Roberts.

2.7.1 Woodward's Counterexample

Woodward has pointed out that the law 'All charged objects accelerate at n m/s^2' qualifies as an acceptable cp-law for *any* n, according to Pietroski and Rey's account. For *any* n, it is possible to find an independently confirmable factor, e.g., electric fields, that explains deviations from the cp-law in question.

In dealing with this case, it is important to clearly distinguish the system from the disturbing factor. Part of the problem is that there are various readings of Woodward's counterexample. One way to read it is to take the charged object as the system in question and the electric fields as the disturbing factors. On the basis of our augmented account of the epistemic acceptability of cp-laws, this reading yields *one* cp-law that is epistemically acceptable: 'All charged objects accelerate at 0 m/s^2.' For any other n the claim is false. Variation of the disturbing factors by either experimental or theoretical means yields the conclusion that in the absence of the disturbing factors (external fields etc.) charged objects do not accelerate. What we get is Newton's first law applied to charged objects.

According to another reading, the system whose behaviour is characterised by the cp-law is not the isolated charged object but rather the *charged object in a field F*. In this case, the disturbing factors are *additional* fields over and above F. Again, if we vary the disturbing factors, i.e., the *additional fields*, we get just one cp-law as a result, not a whole class. In the absence of disturbing factors, the acceleration of the charged object in F will be such and such. Thus, after (1) clearly separating the system under investigation and the disturbing factor and (2) taking on board our refined notion of epistemic acceptability, Woodward's examples cease to be problematic.[14]

What is important in this case, and even more so when it comes to Earman and Roberts' counterexample, is to keep in mind that two aspects of the testing of cp-laws have to be kept separate. First, we need to be able to know what would happen if no disturbing factors were present. For this we need an experimental or theoretical methodology that allows us to vary the disturbing factors. The second aspect of the confirmation/disconfirmation of cp-laws consists in checking the alleged nomological connection. In the case of the charged object in a field F, these two aspects amount to the following. The first aspect consists in varying the *additional* fields in order to find out what happens in their absence. The second aspect consists in finding out how the charged object behaves in the field F, which is considered as a *part* of the system under consideration. This second aspect typically requires the variation of the charge and the field F; thus, it concerns factors that are *internal* to the system – it does not concern external disturbing factors. This second aspect, i.e., the variation of the system-internal factors, will be part of the confirmation of any law, whether cp or strict. What is characteristic of the confirmation of cp-laws is the first aspect: The experimental or theoretical variation of the disturbing factors.

2.7.2   *Earman and Roberts's Counterexample*

Earman and Roberts presented 'If something is spherical it is electrically conducive' as a counterexample to Pietroski and Rey's account. It will always be possible, they argued, to point to the molecular structure as an

---

[14] Wolfgang Spohn (personal communication) raised the following objection to this account: There is surely something wrong with an account of cp-laws that allows both cp(All A are B) and cp(All A are non-B) to be true. On my account, both cp(All humans are brown-eyed) and cp(All humans are blue-eyed) could turn out to be true, provided we can intervene in some of the difference-making genes.

Rejoinder: It might indeed be true that at some point we can intervene in humans so that 'All humans are brown-eyed' or 'All humans are blue-eyed' could come out as true. But that does not suffice for the relevant cp-laws to be both true. What I argued was that the first thing to clarify is what the relevant systems are – the systems with respect to which the cp-law is asserted. When we talk about humans in an unqualified sense we are talking about non-manipulated human beings. With respect to these – on the account presented – we have to find out what is true about them if there are no intervening or disturbing factors. So, we have the claims:

(i)   In the absence of disturbing factors all humans are brown-eyed.
(ii)   In the absence of disturbing factors all humans are blue-eyed.

For all we know, both (i) and (ii) come out as false, in contrast to the initial assumption.

independently confirmable factor that explains why certain round objects fail to be conducive. The example thus meets Pietroski and Rey's conditions, but it does not seem to be an acceptable cp-law.

Here, again, it is important to separate system and disturbing factor. The cp-law says that round objects have the disposition to be electrically conducive. The alleged disturbing factor in this case is the molecular structure of the object. For a cp-law to be epistemically acceptable we have to follow the two-step testing strategy we have outlined. The first step consists in finding out what would happen if the alleged disturbing factor, i.e., the molecular structure, were absent. This, however, is simply not possible in this case. It is neither experimentally nor theoretically possible to determine how a spherical physical object behaves in the absence of its molecular structure. What this shows is that for something to qualify as a disturbing factor for an epistemically acceptable cp-law, it must be (nomologically) possible for the system in question to exist without the disturbing factor. This independence requirement is violated in the example under consideration. Thus, experimental or theoretical methods to find out what would happen in the absence of the disturbing factor cannot be applied. A fortiori, the alleged cp-law fails to be epistemically acceptable.

2.8   Laws, Dispositions and Invariance

In the previous sections I established that law statements attribute dispositions to systems, with dispositions being properties that can be instantiated without being manifest. In accounting for the role law statements play in scientific practice, I did not appeal to specific modal features of dispositions. In this section I will address these features in more detail. It will prove fruitful for the discussion of the modal aspects of dispositions not to approach this issue directly but rather to start by recapitulating the metaphysical claims argued for so far.

2.8.1   Recapitulation

In Chapter 1, I argued that the role of law statements in confirmation and testing is best understood by assuming that law statements attribute some specific behaviour to *systems* – where a system is broadly construed as being any kind of entity or object in whose behaviour we are interested, ranging from elementary particles to populations of fruit flies, economies or spacetimes.

Next, I argued that the use of law statements in explanation, prediction and manipulation can be best made sense of if we assume what I called the 'modal surface structure' of laws. The fact that internal generalisations come with a domain of quantification is best accounted for by assuming that law statements attribute a space of possible states to systems. Internal generalisations, on their own, do not tell us which state a system is *actually* in. The domain of quantification for the internal generalisations gives us *purely modal information*: some states are assumed as being possible states for the system.

Besides attributing a space of possible states, the internal generalisations by virtue of the law equations *restrict* these states as well as the temporal evolutions of the systems. As a consequence, the role that internal generalisations and in particular law equations play in scientific practice is best understood by assuming not only that they describe relations between variables but also that these relations work as *constraints*. I furthermore argued that the modal element in 'restriction' and 'constraint', which is usually explicated in terms of nomological necessity, is best understood in terms of invariance relations. I distinguished the following respects of what is usually assumed by the inviolability of laws or their nomological necessity:

1) Invariance of the law equation with respect to initial conditions: The law equation holds irrespective of which values from a certain range of initial conditions, or boundary conditions or other sets of variables, obtain or would obtain.
2) Invariance of the law equation with respect to other features of the system: In the case of macroscopic laws, two kinds of properties with respect to which invariance may occur ought to be distinguished:
   a. Same-level properties,
   b. Lower-level or constitutional properties.
3) Invariance of the law equation with respect to the behaviour of other systems in the universe.

Furthermore, I showed that even though invariance with respect to actual and counterfactual changes is a modal notion, it is a scientifically accessible (testable) relation. Finally, I argued that a minimal metaphysics of scientific practice – given that it relies on inference to the best explanation of the success of characteristics of scientific practice in order to establish metaphysical claims – should refrain from speculating on whether or not the modal surface structure can be reduced to essences, dispositions or non-modal facts. Such hypotheses do not contribute anything to understanding how laws are used in explanation, prediction or manipulation.

In Chapter 2, I argued that the properties law statements attribute to systems are best understood as dispositional properties. This is due to the fact that law statements are often attributed to systems that are immersed in the universe and thus a part of it (rather than the universe as a whole). It follows that other systems (other parts of the universe) may interact with the system in question. As a consequence, the law equation attributed to the system (and appealed to in explanations) may fail to describe the behaviour of the system in question due to interferences. The behaviour of the system is thus not invariant with respect to actual or counterfactual changes in the behaviour of systems in the environment. However, the law statement is nevertheless taken to characterise the system in question even if actual or counterfactual changes obtain. This practice is best understood if law statements are read as attributing properties to systems that can be instantiated but may fail to be manifest, thus reconstructing law statements as ascriptions of dispositions.[15]

2.8.2   Dispositions and Modality

In several accounts of laws of nature, the modal aspects of dispositions have played an explanatory role for the modal status of laws. Thus, Bird holds:

> If properties have a dispositional essence then certain relations will hold of necessity between the relevant universals; these relations we will identify with the laws of nature. [...] Since the relevant relations hold necessarily this view is committed to necessarianism about laws – laws are metaphysically necessary. (Bird 2007, 43)

Anjum and Mumford write:

> In the sense of natural possibility, therefore, it would actually be the case that if F is possible then there is a disposition towards F. No contrast is being offered, therefore, between dispositionality and possibility. On the contrary, the former may ground entirely what the latter consists in. What better candidate could there be for supplying the world with natural possibility than the dispositions that particular things have? (Mumford and Anjum 2011, 182)

---

[15] This analysis has a further consequence. Because the law statement attributes a property (a dispositional property, to be more precise) to a system and the system instantiates this property independently of changes in other systems of the universe, the disposition qualifies – *pace* Ladyman and Ross – as an *intrinsic* property of the system in question (see Section 1.1.4 for further discussion).

These are two attempts to explicate natural modalities in terms of primitive modal features associated with dispositions. In Bird's case it is the fact that dispositions constitute the essence of properties that plays an explanatory role, while in the case of Anjum and Mumford it is the modal relation between a disposition and its manifestation that is taken to be primitive and explanatory.

It is important to keep these two approaches separate, because they appeal to different modal relations. First, there is the issue of the modal relation between a system and the disposition that the law statement attributes to it. According to Bird, an electron has its dispositional properties essentially. The second issue is the display or manifestation of the disposition (which may or may not be an essential disposition of the system). 'The connection has modal strength in that it provides the world with more than pure contingency' (Mumford and Anjum 2011, 175). What can I say about these two modal features of dispositions?

In Chapter 1, I explicated the modal aspects of law statements in terms of invariance relations. In accordance with the approach of a minimal metaphysics for scientific practice, I refrained from speculating about the underpinnings of the invariance relations. Thus, rather than postulating a metaphysical base for invariance relations in terms of modal features of dispositions, I suggest, conversely, understanding the modal features of dispositions as far as possible in terms of scientifically accessible invariance relations. This is part of the project to account for all the natural modalities we encounter in scientific practice in terms of invariance relations.

Let me start with what I called the first issue – the relation between a system and the system's disposition. In the previous sections of this chapter I argued that the use of ceteris paribus clauses and the practice of extrapolation is best understood by reading law statements as attributing dispositions to systems. In other words, the examination of the practice of extrapolation shows us how to understand the invariance of the law with respect to actual and counterfactual changes of other systems. It is the disposition to behave according to the law equation that remains invariant rather than the overt behaviour. But once we have established that having the disposition is invariant with respect to actual and counterfactual changes in other systems, it seems to me that we have accounted for why and in what sense we conceive of the relation between a system and its disposition as necessary or even essential: There are no actual or counterfactual changes that may bring about the system not having the disposition.

Dealing with the second issue is a bit trickier, but I will argue that the same strategy as with the first issue works here as well. What is sometimes called 'dispositional modality' can be explicated in terms of (conditional) invariance relations.

The issue under consideration concerns the relation between a system instantiating a disposition plus the stimulus conditions (if there are any), on the one hand, and the display of the disposition, on the other hand. This relation is typically described in modal terms. According to Tugby, for instance, a disposition is 'oriented' towards its manifestation (Tugby 2013, 452). 'The connection has modal strength in that it provides the world with more than pure contingency' (Mumford and Anjum 2011, 175).

Given our analysis of the modal surface structure, we can explain why the relation between a system instantiating a dispositional property and the manifestation of this property is not merely accidental but obtains with some sort of necessity. So, again, what I suggest is the converse of the programme of Bird or Anjum and Mumford: Given the notion of invariance as a scientifically accessible modal notion, we should invoke invariance relations to give an account of what might be meant by dispositional modality.

The display or manifestation of a disposition sometimes fails to be invariant with respect to changes in other systems – that is why some law statements need to be hedged by ceteris paribus clauses. As a consequence, we cannot say that given a system instantiates a disposition D and all the stimulus conditions S (if any) are present, the manifestation M will occur, whatever the behaviour of other systems in the universe. The discussion of antidotes and interfering factors has shown that this would be the wrong model.

So, what can we say? Let us have a look at Newton's first law again:

> Every body continues in its state of rest or of uniform motion in a straight line, unless it is compelled to change that state by forces impressed upon it. (Newton 1999, 416)

The behaviour of the bodies is not invariant under *all* changes in the environment. There are two kinds of changes. If the changes give rise to impressed force on the body in question, the disposition will fail to be (completely) manifest. However, as long as the changes are such that they do not give rise to impressed forces, the disposition will be manifest. That is, the manifestation of the disposition is invariant under the second kind of changes but not under the first kind of changes. What I suggest is thus that what Anjum and Mumford characterise as a '*sui generis* modality that is less than necessity but more than pure contingency' (Mumford and Anjum 2011, 85) can be explicated in terms of a conditional invariance relation: In the case of Newton's first law, if there are no impressed forces, every body continues in its state of rest or of uniform motion in a straight line – whether or not there are any other (i.e., not force-inducing) actual or counterfactual changes in the behaviour of other systems in the universe.

But what if there are impressed forces or interfering factors? This question brings us back to considering the multitude of cases that might obtain conditional on the presence of various interfering factors. Recall the list of particularised regularities we discussed in Section 2.4.3:

(i) $\forall x$ (Cx & no other interfering factor $\supset M_0x$)
(ii) $\forall x$ (Cx & I_1x & no other interfering factor $\supset M_1x$)
(iii) $\forall x$ (Cx & I_2x & no other interfering factor $\supset M_2x$)
(iv) $\forall x$ (Cx & I_3x & no other interfering factor $\supset M_3x$)
. . .

(. . .) $\forall x$ (Cx & I_1x & I_2x no other interfering factor $\supset M_{12}x$)
. . .

where Cx stands for system x being in circumstances C, I_ix for x being affected by an interfering factor and M_ix for x displaying behaviour M.

For every regularity listed, it is true that if the antecedent condition is satisfied, then whatever actual or counterfactual further changes occur with respect to the behaviour of other systems in the universe, the manifestation referred to in the consequent will occur. So, given the antecedent, the occurrence of the manifestation is invariant with respect to further changes.

It may be objected that this claim is trivially true because if the manifestation has not occurred the changes must have been interfering changes. Two things need to be said here. First, the problem is at most a problem about how we find out whether a change is an interfering or a non-interfering change – not a problem about the distinction in itself. Second, in the case of a regularity such as (ii) $\forall x$ (Cx & I_1x & no other interfering factor $\supset M_1x$), it is not trivial that for all x, M_1x occurs if the Cx & I_1x is satisfied and there are no further interfering factors. It is due to laws of composition that what is referred to in the antecedent guarantees M_1 to occur. It is ultimately the invariance of the laws of composition that ensures that if a number of factors contribute to the display of a behaviour, a determinate display occurs invariantly. So, the relation between a disposition and its manifestation can be explicated in terms of invariance relations that are conditional on the presence of interfering factors.

To conclude this section, neither in Chapter 1 nor in Chapter 2 did we appeal to modal features of dispositions to explain why we have the scientific practice we have. Modal features of dispositions thus play no primitive role in our metaphysics. On the contrary, we showed in this section how to explicate modal features of dispositions in terms of invariance relations.[16] This is, as mentioned before, part of the project of accounting for all the natural modalities we encounter in scientific practice in terms of invariance relations. No further dependence relations need to be postulated at this stage. It might, of course, be objected that the question 'Why does invariance hold?' has not been answered. Such answers might rely on essences, the Humean mosaic, etc. It might, however, also be argued that we should take the job of the sciences to be to uncover primitive invariance relations. Be that as it may, such accounts do nothing to explain the success of features of scientific practice. So, we should be content with invariance relations.

2.9   Comparisons with Other Dispositional Accounts of Laws

Let me finally compare the view developed here with other dispositionalist accounts of laws. First, by way of recapitulation, in positing dispositions I have not committed myself to any form of essentialism (such as, for instance, Ellis 2001, Bird 2007 or Chakravartty 2007) or to the claim that dispositions should be understood as causal powers (such as, for instance, Ellis 2001, Bird 2007, Chakravartty 2007 or Anjum and Mumford 2011). Furthermore, I am not committed to the view that a disposition *needs* a stimulus condition to become manifest; the absence of interfering factors might suffice for manifestation. The only positive features I appealed to in the argument from extrapolation were the following: Dispositions are properties, they may be multi-track, and they can be distinguished from categorical properties because they allow a distinction to be drawn between a property being instantiated and a property being manifest. Additionally, I just made a suggestion on how to understand the modality that characterises the relation between a system having a disposition and the occurrence of a stimulus condition (if necessary), on the one hand, and the

---

[16] It might be objected that the fact that a particular disposition has a particular manifestation should be viewed as metaphysically necessary. In other words, dispositions might be viewed as having irreducible modal features/aspects that are not automatically bound up with invariances. The rejoinder depends on the view one takes concerning the individuation of dispositions. If a disposition is individuated or defined in terms of its manifestation, it is indeed necessary that it has a particular manifestation, but it appears to be doubtful that this necessity is a natural necessity. If, however, the individuation of the disposition depends on other means, e.g., empirical means, the question of whether or not the disposition comes with a particular manifestation is presumably a case of a natural modality – but then it seems that it can be spelled out in terms of invariance relations. (Thanks to an anonymous referee for raising this issue.)

manifestation of the disposition, on the other, in terms of invariance relations.

Differences among the various dispositional accounts of laws of nature do not only concern the conception of a disposition. Further differences – and I will confine myself now to contrasting Bird's view with mine – become apparent if we examine the exact content of the law statements that are considered. Such differences can be illustrated by an example, the two-body law (TBL). Let us consider a system consisting of two masses M and m at positions R and r plus Newtonian gravitational force $(-GMm/(R - r)^2)$ between them. According to my proposal, the relevant law is something like this:

(TBL) All massive, gravitationally interacting two-body systems are disposed to behave according to the Langrangian equations with the gravitational force $F = -GMm/(R - r)^2$.

On my account, the two-body system has a disposition, namely, to behave according to the Langrangian equations with the force $F = -GMm/(R - r)^2$ (or the corresponding potential). The disposition becomes manifest if there are no interfering factors. It is a multi-track disposition; it has tracks for various Ms and ms as well as for different Rs and rs.

By contrast, Bird, following Shoemaker's views on the identity of properties, would be interested in the property M (see Bird 2007, 22). He would consider a disposition D_x whose stimulus is a mass m_x at a displacement $R - r$ and whose manifestation is the exertion of the force $F_x = -GMm_x/(R - r)^2_x$. Here x serves as an index for a particular value of the variable set of values for the variables that give rise to a single-track disposition D_x. The disposition D_x is part of what constitutes the essence of mass M. M has countless dispositions for specific values of m and other distances that are all constitutive of M's essence. All these give rise to separate laws, one each for each disposition.

So, there are two additional and significant differences compared with my account. The first concerns the bearer of a disposition. In the case of TBL, what Bird is interested in is not a disposition of the compound two-body system but, rather, dispositions that characterise properties of the parts of the systems. It is the fact that these dispositions are essential dispositions that accounts for the necessity of laws. By contrast, my starting point is the system whose behaviour we want to characterise, i.e., the two-body system. On my account, TBL attributes a disposition to two-body systems.[17]

---

[17] I take the fact that the disposition is attributed to a compound system to be rather uncontroversial. Many of the dispositions we are best acquainted with are dispositions of macroscopic objects, which are compounds.

The second difference concerns tracks of dispositions. We have seen before that a bundle of single-track dispositions would not be able to account for the practice of extrapolation (Section 2.4.3). Thus, on my account, TBL attributes one multi-track disposition to the system. By contrast, Bird insists on a multitude of single-track dispositions.

It is important to point out these differences because objections to Bird's view are not necessarily objections to my view. This can be illustrated by the case of a major objection that has been raised against dispositional accounts of laws of nature. The issue in question is that of symmetries and conservation laws (for a discussion see Livanios 2010 and French 2014, 249–54). Bird himself acknowledges the challenge:

> Several of our most important laws state that certain quantities are con-served in all interactions: mass-energy, charge, momentum. Lepton num-ber, angular momentum, etc. Corresponding to these are laws asserting that the universe displays certain symmetries. It is difficult to see why, for example, when two charged objects interact, it is a manifestation of a dispositional essence that the total charge should remain constant. [...] In any case, there is something mysterious about conservation laws [...] How does a system know that energy should be conserved? (Bird 2007, 213)

Bird is interested in explaining constraints on the behaviour of sys-tems in terms of the essences of their properties, e.g., in terms of dispositions like D_x, which is had by M essentially. Additional con-straints by conservation laws have no place. 'Properties are already constrained by their own essences and so there is neither need nor opportunity for higher-order properties to direct which relations they can engage in' (Bird 2007, 214).

This problem for the dispositional essentialist does not arise for me because I am not attempting to explain constraints on the behaviour of systems in terms of the essences of properties. Rather, I take laws to be describing how systems behave and to what extent this behaviour is invariant with respect to certain changes. What invariances there are is, on my account, an empirical matter. Thus, it may simply turn out that the behaviour as described by the law predicate is invariant with respect to certain transformations, e.g., translations in space and time. In the example of TBL, the law statement attributes to the two-body system that it is disposed to behave according to the Lagrange equations with a gravitational potential. Since this potential is independent of time, energy is conserved. As a consequence, the two-body systems are disposed to a behaviour that conserves energy.

Since I do not attempt to provide a metaphysical underpinning for invariance relations, I don't have to deal with the question of whether essences can account for them. As I argued before, reducing the modal notion of an invariance relation to essences, the Humean mosaic and so on is beyond the scope of a minimal metaphysics of scientific practice.

What this discussion shows is that there are two very different strands of argumentation that led to the discussion of dispositions in metaphysics of science – one via Mill, Cartwright and the extrapolation argument, and another via Shoemaker and considerations of the identity of properties. While the focus of the former argument is to generate an understanding of aspects of scientific practice, the focus of the latter is to try to give an account of what nomological modality is grounded in. These considerations are very different. Authors who commit themselves to one line of reasoning for dispositions need not commit themselves to the other. And arguments against one of these lines of reasoning do not automatically transfer to the other.

CHAPTER 3

Causation – Conceptual Groundwork

In this chapter, I will analyse the practice of causal explanation. I am in particular interested in whether additional metaphysical assumptions – over and above those discussed in Chapters 1 and 2 – are necessary to understand our causal practice. The claim I will defend is that no additional assumptions are required. While this claim on its own may not be particularly controversial, I intend to provide a novel account of how to fit causation into the world as described by the sciences and their laws.

The practice of causal reasoning or causal explanation in (applied) science may concern either causal generalisations ('smoking causes lung cancer') or singular causal statements ('The bridge collapsed because of poor maintenance'). In what follows I will argue that singular causal statements are a good starting point to develop an account of causation that makes no further metaphysical assumptions over and above those discussed in the previous chapters.

More particularly, I will argue that our practice of identifying token causes (or actual causes) can be fully explained in terms of what I call 'quasi-inertial behaviour' and interfering factors. This quasi-inertial behaviour can in turn be explicated in terms of the (multi-track) dispositions that law statements attribute to systems, which were discussed in Chapter 2. This reduction of macroscopic causal relations to facts about quasi-inertial behaviour and thus dispositions of systems implies that no additional metaphysical structure needs to be postulated to account for the role causation plays in scientific and everyday contexts. Chapter 3 will provide the general outline of how to account for causal claims in terms of the metaphysics outlined in the previous chapters. Chapter 4 will put the account to work, augment it in the light of challenges and explain why it tracks the causal intuitions we have.

3.1 Introduction

For quite a while the ultimate aim of theories of causation was to provide necessary and sufficient conditions for 'the' concept of cause. More recently a number of authors have given up on this project. Ned Hall in a well-known paper (2004) argues that there are two concepts of causation. Nancy Cartwright claims: 'The important thing is that there is no single interesting characterizing feature of causation' (Cartwright 2007, 2). As a consequence, *causal pluralism* has gained some currency: 'Causal Pluralism is the view that causation is not a single kind of relation or connection between things in the world. Instead, the apparently simple and univocal term "cause" is seen as masking an underlying diversity' (Godfrey-Smith 2009, 326).

The observation that there is more than one concept of causation seems correct to me, as I will argue later. However, it does not suffice to merely state this observation; it should be developed into an account of how these different concepts are related. Wittgenstein's concept of family resemblance might be a natural starting point. I do, however, think that a stronger claim can be made: In what follows I would like to argue that there is a *focal concept* of causation. This is an idea that Aristotle introduced in his *Metaphysics* in connection with his envisaged science of *being*. Aristotle observed:

> There are several senses in which a thing may be said to be, as we pointed out previously in our book on the various senses of words; for in one sense it means what a thing is or a 'this', and in another sense it means that a thing is of a certain quality or quantity or has some such predicate asserted of it. (Aristotle 1984, 10–14)

This plurality of concepts of being seems to suggest that there cannot be a single science of being. However, Aristotle then goes on to argue that the plurality of concepts does not entail that no unified account of being can be given. As an analogy, he discusses the case of health. An individual person, an apple and a walk may all be healthy, though in different senses of the word. However,

> Everything which is healthy is related to health, one thing in the sense that it preserves health, another in the sense that it produces it, another in the sense that it is a symptom of it, another because it is capable of it. (Aristotle 1984, 35–7)

The example nicely illustrates what has been called the theory of focal meaning: there are different concepts (of health or of being) but they are all related (in various ways) to one central concept. Similarly, in what follows I will argue that different concepts of causation are not merely related by family resemblance but rather that there is a focal concept of causation.

In the bulk of this chapter I will explicate this focal concept in terms of quasi-inertial processes (i.e., processes systems are disposed to provided there are no interfering factors) and interfering factors (Sections 3.4 and 3.5). The focal concept, I will argue, is of particular philosophical importance for two reasons. First, it is the conceptual bedrock for concepts of causation that explicitly figure in scientific causal explanation, e.g., probabilistic concepts (see Section 3.8) or causal reasoning based on structural equations (see Section 4.5). Second, it tracks many of our causal intuitions.

The relevance of the proposed account of causation for metaphysics consists in the way these concepts of causation are related to science: I argue that the truth-makers for causal claims (for the focal concept as well as for the other senses) do not require anything over and above ordinary scientific facts. As a consequence, the metaphysical assumptions I introduced in Chapters 1 and 2 – systems being disposed to a certain kind of behaviour – suffice to account for the use we make of causal terminology. No additional metaphysical assumptions have to be introduced to account for the role of causation in scientific practice.

Historically, however, it has been doubted that causation fits into a world as described by the sciences. I will start by briefly recapitulating the arguments that were discussed by Russell, Mach and others.

3.2   Historical Background

In 1876 Gustav Kirchhoff criticised the notion of causation in the well-known preface to his *Lectures on Mechanics*:

> It is common to define mechanics as the science of forces and forces as causes that bring about motions or tend to bring them about. [...]. [This defin-ition] is tainted by an obscurity that is due to the notion of cause and tendency [...]. For this reason I propose that the aim of mechanics is to describe the movements that take place in nature – more specifically to describe them completely and as simply as possible. What I want to say is that we should aim at stating what phenomena there are rather than to determine their causes. (Kirchhoff 1876, preface)

Kirchhoff's criticism concerns causation in a productive sense or under-stood as a force, which was indeed the main concept of cause that was used by physicists in the first half of the nineteenth century. Eliminating productive causes from physics still leaves room for other conceptions. Gustav Theodor Fechner and the early Ernst Mach, for example, turned to John Stuart Mill. Mill had repudiated productive causes and defended a regularity view according to which the cause is an instance of the antecedent in a law, which is a sufficient condition for the occurrence of the effect: 'To certain facts certain facts always do, and, as we believe, will continue to, succeed. The invariable antecedent is termed the cause; the invariable consequent the effect' (Mill 1974, Vol VII, 327). When at the turn of the century Ernst Mach and Bertrand Russell criticised the notion of cause, it was a Millian regularity view of causation they had in mind.

Mach argued that, for at least three reasons, the concept of cause cannot be applied to reality as described by physics and should therefore be given up.

(i) Taking seriously the concept of cause as a set of conditions implies that you need to consider every single factor on which an event depends. That is practically impossible.

> If one attempts to eliminate the traces of fetishism, which are still associated with the concept of cause, and takes into consideration that in general you cannot specify a cause since a phenomenon most of the times is determined by a whole system of conditions, you are led to the conclusion to give up the concept of cause altogether. (Mach 1900, 435–6)

(ii) The concept of cause requires strict regularities. However, there are no such regularities:

> In nature there are no causes and no effects. Nature is there only once. Repetitions of identical cases, such that A is always correlated with B, the same outcome under the same conditions, i.e. what's essential for the relation between cause and effect, exists only in abstraction [. . .]. (Mach 1982, 459)

(iii) Finally, the advanced sciences replace causal terminology by the concept of a mathematical function, which is a more precise notion.

> In the higher developed natural sciences the use of the concepts cause and effect becomes more and more limited. There is a perfectly good reason for this, namely that these concepts describe a state of affairs provisionally and incompletely – they are imprecise [. . .]. As soon as it is possible to charac-terise the elements of events through measurable quantities, [. . .] the dependence of the elements on each other can be characterised more completely and more precisely through the concept of a function, rather than through the insufficiently determined concepts cause and effect. (Mach 1982, 278)

Russell in his well-known paper 'On the Notion of Cause' adds two important considerations:

(iv)   Causes are usually conceived as localised events (locality). However, no localised event is sufficient for the occurrence of any other event because there may always be an interfering factor.

> In order to be sure of the expected effect, we must know that there is nothing in the environment to interfere with it. But this means that the supposed cause is not, by itself, adequate to insure the effect. And as soon as we include the environment, the probability of repetition is diminished, until at last, when the whole environment is included, the probability becomes *nil.* (Russell 1912/13, 7–8)

Thus, if the fact that causes determine their effects is spelled out in terms of conditional regularities such that the cause is the antecedent, we cannot have both locality and determination.

(v)   In a physical system as described by fundamental physics, the future determines the past in exactly the same way as vice versa.

> [...] the future 'determines' the past in exactly the same sense in which the past 'determines' the future. The word 'determine', here, has a purely logical significance: a certain number of variables 'determine' another variable if that variable is a function of them. (Russell 1912/13, 15)

Therefore, the asymmetry that we associate with the causal relation is inconsistent with fundamental physics.

   Mach and Russell both conclude that there is no place for causation in the advanced sciences. Even though the claim that today's advanced sciences no longer use the term 'cause' has been proven false,[1] the question has been raised of how to reconcile Mach's and Russell's observations with the persistence and usefulness of causal terminology. I will return to this question in Section 3.9.

3.3   The Focal Concept of Causation: Overview

The aim of this chapter is to explain how what we usually mean by 'causation' (in the sense of 'token causation' or 'actual causation') fits into a world as described by the sciences in terms of laws of nature or other regularities. This is important because – if successful – the account of causal reasoning and causal explanation need not make any metaphysical

---

[1]   See, for instance, Suppes (1970), 5–6 and Williamson (2009), 195–7.

assumptions to account for the practice of causal explanation over and above those introduced in Chapters 1 and 2.

The relevant concept of cause – what I will call the 'disruptive' concept of cause – will be demonstrated to be a focal concept of cause by showing how other concepts are related to it.

The disruptive concept of cause is central for three reasons: First, it is central to an understanding of why causal terminology plays an important role in scientific as well as in everyday contexts; second, with this concept as our starting point, we can understand other uses of 'cause'; and third, as a bonus, it turns out that it is the concept that is tracked by many of our causal intuitions.

In Sections 3.4 and 3.5, I will start to explicate the disruptive concept of causation. This will comprise two steps. The first step will analyse causation in terms of quasi-inertial processes and interferences. The second step will show that quasi-inertial processes and interferences can be understood in scientific terms, i.e., in terms of dispositions of systems.

3.4   The Disruptive Concept of Causation: First Step of an Analysis

Mach observed at the beginning of the last century that '[i]n general we only feel the need to ask for a cause, if a (unexpected) change has occurred' (Mach 1900, 432; for similar observations see Kahneman and Miller 1986, 148; Hitchcock and Knobe 2009). While this remark expresses a psychological observation, Hart and Honoré have similarly analysed what they take to be our ordinary common-sense concept of causation. They observe that '[c]ommon experience teaches us that, left to themselves, the things we manipulate, since they have a "nature" or characteristic way of behaving, would persist in states or exhibit changes different from those which we have learnt to bring about in them by our manipulation.' They then go on to state:

> The notion, that a cause is essentially something which interferes with or intervenes in the course of events which would normally take place, is central to the common-sense concept of cause. (Hart and Honoré 1959, 27)[2]

---

[2] In contrast to Hart and Honoré, I will not discuss quasi-inertial behaviour in terms of 'normality' but rather in terms of quasi-inertial behaviour, which might be taken to be a different way of spelling out a system's 'characteristic way of behaving'.

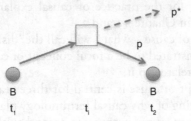

p: B's actual path
p*: the path B would have taken if A had not interfered at time t$_i$

Figure 1 Deflection of billiard ball

These psychological and semantic observations suggest an account of causation according to which what we pick out by the word 'cause' is an interference with a quasi-inertial process. In what follows I will sketch such an account.[3]

Let me introduce the central idea by way of a simple and idealised example (Figure 1). Two billiard balls, A and B, bounce against each other at t_i and are deflected. We take the presence of ball A at a particular place at t_i as the (actual) cause for B's deflection.

What makes our causal judgment correct? According to Newton's first law, B will display a certain inertial behaviour, namely, to simply continue in a straight line with uniform motion (path p*) provided there are no interfering factors. Newton's first law describes the (quasi-)inertial behaviour of B. (I take *quasi-inertial behaviour* to be a generalisation of *inertial behaviour* that refers to the temporally extended behaviour of systems that are not interfered with; see Section 3.5.1 for details). However, this quasi-inertial behaviour of B is not displayed; B takes path p rather than path p*. If the quasi-inertial behaviour does not occur, Newton's first law tells us that there must be some factor that interacted with B (the box in Figure 1 represents the interaction). The event of A interacting with B is the cause of the quasi-inertial behaviour not being displayed. More generally, the cause is something that interacts

---

[3] The term 'bring about' in one of these quotations might suggest that common sense or science takes *production* as the focal concept. However, given that the practice of causal reasoning can be explained in terms of interferences into quasi-inertial processes, for a minimal metaphysics of scientific practice there is no need to invoke a potentially metaphysically richer notion such as *production*. (Thanks to an anonymous referee for raising this question.)

with or disturbs the system under consideration such that instead of the latter's quasi-inertial behaviour, another behaviour occurs.

More precisely and translated into event talk, this can be stated as follows ('DC' stands for the definition of the disruptive concept of causation):

Let Z_i (t_2) be the state that a system S (at an earlier time t_1) is disposed to be in – provided nothing interferes between t_1 and t_2. Let e be the event of S being in state $Z_e(t_2)$ at t_2, where $Z_e(t_2) \neq Z_i(t_2)$.

(DC) An event c is a cause$_{DC}$ of an event e iff c is the event of some factor interfering with the quasi-inertial behaviour of system S between t_1 and t_2 such that S at t_2 is in state Z_e rather than in state Z_i.[4]

(DC) gives us a contrastive notion of causation because the cause is a cause for S being in state Z_e *rather than in state Z_i*. (DC) is contrastive with respect to the effect (cf. Schaffer 2005, who advocates contrastivity with respect to causes as well. (DC) can easily be made contrastive with respect to causes as well if the occurrence of c is contrasted with the absence of c.)

Clearly, there are a number of terms in (DC) that need to be explicated; 'quasi-inertial behaviour' and 'interfering factor' will be treated in Section 3.5. The phrase 'such that' will be dealt with in Chapter 4. It will turn out that it does not suffice to merely identify the interferences; we furthermore have to make a distinction between those interferences that are relevant and those that are not. Those that are *relevant* are interferences *such that* the effect occurs.

Let me, however, start with another illustration of what is involved in a causal claim according to this first step of analysis. In this account we should analyse a claim like 'The striking of the stone causes the shattering of the window at t' as being true in virtue of the following:

(1)   An event e (the effect) obtains: System S (the window) is in state Z_e at t (it is shattered).
(2)   The quasi-inertial behaviour of S would have led to S being intact at t (S in state Z_i, which is different from Z_e): In the absence of interfering factors, the window, which as a matter of fact is broken, would have remained intact.
(3)   An event c obtains (the stone hits the window), which is an interference with the quasi-inertial process such that S at t is in state Z_e rather than Z_i.

---

[4] As indicated at the outset, my aim in this and the following sections is not to cover all uses of the word 'cause' but rather what I have called the 'disruptive' concept of cause. Other concepts will be dealt with in Section 3.12.

What is essential for the determination of the quasi-inertial behaviour or quasi-inertial process – as will become clear in the following discussion – is that we start with the effect in question. According to the account presented here, the effect consists of a system S being in a state Z_e at a certain time t that is different from or deviant relative to the quasi-inertial behaviour of the system at that time. When looking for a cause, we are looking for something that brought about this deviation.

3.5   The Disruptive Concept of Causation: Second Step of an Analysis

The second step of the analysis of the disruptive concept of causation consists in showing that the crucial notions in the characterisation of actual causation ('quasi-inertial process' and 'interfering factor') can be understood in scientific terms, that is, in terms of dispositions of systems. What is essential for the account to work is not that all quasi-inertial processes have a common physical feature, but rather (1) that in every single case of actual causation there is a determinate and objective fact of what the quasi-inertial process of the relevant system is and (2) that in every single case the quasi-inertial behaviour can be characterised in terms of laws, so as to ensure that causal facts and causal dependence can be explicated in terms of those metaphysical assumptions introduced in the previous chapters.

3.5.1   Quasi-Inertial Processes

Central to my account of causation is the notion of a *quasi-inertial process*. In physics, the *inertial motion* of a massive particle is defined as the motion of a free particle, i.e., a particle free from external forces acting on it. My notion of *quasi-inertial process* or *quasi-inertial behaviour* generalises this aspect of inertial motions: it refers to the temporally extended behaviour of systems that are not interfered with. A few examples should provide us with a better grip on this notion.

Example 1: Newton's first law describes the (quasi-)inertial behaviour of a massive particle: 'Every body continues in its state of rest or of uniform motion in a straight line, unless it is compelled to change that state by forces impressed upon it' (Newton 1999, 416). The law describes the systems in question ('every body') that are disposed to display a certain behaviour ('continues in its state of rest or of uniform motion in a straight line') provided there are no interfering factors ('unless it is compelled to change that state by forces impressed upon it').

Example 2: Galileo's law of free fall defines a quasi-inertial process: A free-falling object in a vacuum displays a certain behaviour as long as the falling object is free, i.e., as long as no interfering factors intervene.

Example 3: The Lotka–Volterra equations describe the temporal development of a biological system consisting of two populations of different species: one predator, one prey. The relevant equations for interacting prey–predator populations are (1) $dx/dt = x (a - by)$ and (2) $dy/dt = -y (c - gx)$, where x represents the number of prey and y the number of predators of some kind, and a, b, c and g are constants. These equations describe a quasi-inertial behaviour in that they describe the behaviour that a biological system is disposed to display provided there are no interfering factors (e.g., additional predators or severe floods).

Example 4: According to the economic law of supply and demand, if all other factors remain equal, the higher the price of a good, the less demand there will be for the good. Thus, certain kinds of systems (economies) are disposed to display a certain temporal development provided there are no interfering factors (e.g., state interventions that fix prices).

A number of points should be noted. First, whether a certain kind of system is disposed to display some kind of quasi-inertial behaviour is an objective matter for which the usual sorts of scientific evidence are available. Some claims about quasi-inertial processes can be (more or less directly) empirically tested (e.g., Galileo's law of free fall or the Lotka–Volterra equations), while in other cases, empirical and theoretical considerations together provide warrant for our acceptance of the laws, equations and so on that represent attributions of quasi-inertial processes, for instance in the case of the law of supply and demand. (I have dealt with this issue in Section 2.6 'Testing cp-Laws'.) What counts as a quasi-inertial process and a deviation therefrom may also be constrained by very general theoretical considerations, such as those that led people to give up an Aristotelian conception of the natural world and adopt a Newtonian one instead (and with it different conceptions of quasi-inertial behaviour). The essential point is that we have the usual scientific evidence for characterising quasi-inertial processes. Physics tells us that free-falling objects fall (roughly) according to $s = \frac{1}{2} gt^2$; that is the quasi-inertial process for such objects. It is *false* that free-falling objects fall (roughly) according to $s = \frac{1}{2} gt^4$. This is similar for other disciplines or branches of science; for example, biology tells us that the relevant equations for interacting prey–predator populations are (1) $dx/dt = x (a - by)$ and (2) $dy/dt = -y (c - gx)$, and according to the evidence we have, the equations (1*) $dx/dt = x^2 (a - by)$ and (2*) $dy/dt = -y^2 (c - gx)$ provide *false* descriptions of the temporal evolution of such populations.

Second, the quasi-inertial processes we are looking at might be internally quite complex. Within the quasi-inertial temporal development as described by the Lotka–Volterra equations, there might be all kinds of other processes taking place (e.g., rabbits eating grass or playing). These processes might themselves be quasi-inertial processes of systems that are part of the biological system described by the Lotka–Volterra equations.

Third, the notion of a quasi-inertial process is closely connected to that of an *exclusive* ceteris paribus law (see Section 2.1). Exclusive ceteris paribus laws describe a relation between properties of a system or the behaviour of a system that obtains *provided certain factors are absent*. All four cases mentioned here are examples of exclusive ceteris paribus laws. (Examples 3 and 4 are mixed in the sense that they are also comparative ceteris paribus laws, where *comparative* ceteris paribus laws assume that *factors not explicitly mentioned remain constant*; see Schurz 2002 for this classification of ceteris paribus laws).

Finally, even though I have introduced quasi-inertial processes via ceteris paribus laws so as to make clear that such processes can be fully accounted for by the sciences, I do not claim that there is an explicit ceteris paribus law for every quasi-inertial process. On the contrary: Consider again the example of a falling stone. Galileo's law – a ceteris paribus law – describes one particular quasi-inertial process, namely, free fall. There are, however, many other quasi-inertial processes in the vicinity, e.g., a falling stone in air, water or other media, even though we do not have explicit ceteris paribus laws for the behaviour of these systems. This is an important point, for otherwise one might be tempted to think that the analysis of disruptive causation only works for the idealised cases described in explicit ceteris paribus laws.

3.5.2   *Quasi-Inertial Behaviour, Multi-Track Dispositions and the Relativity of System Individuation*

As we have seen in the previous chapter, systems can be disposed to many different kinds of behaviour depending on the presence or absence of other systems that might work as interfering factors. Newton's first law describes what a system is disposed to provided there are no interfering factors. Newton's second law describes what a system is disposed to provided there are other systems that interact with the system in question. In Chapter 2 this was acknowledged by way of introducing multi-track dispositions. To use a further illustration: Galileo's law describes the behaviour of a falling stone in a vacuum. That is one track of the stone's multi-track disposition (the 'ideal' track). In the presence of a medium M_I with the viscosity V_I, the stone will fall with a different velocity, and similarly for other media and viscosities. These are further tracks of the stone's multi-track disposition. As an equivalent way of characterising the dependence of the behaviour on the presence or absence of interfering factors, we introduced the distinction between complete and partial manifestation (see Section 2.4.3). Instead of describing the system's conditional behaviour in terms of a multi-track disposition, we can then say that the system's disposition is completely manifest in the absence of interfering factors (corresponds to the ideal track) and partially manifest in the presence of interfering factors – that is, the disposition contributes to the behaviour of the compound consisting of the stone and the medium in question. In other words, the various tracks of the multi-track disposition correspond to the complete manifestation and to different ways of partial manifestation.

What is important for the following analysis of causation is the fact that there is a correspondence between the different tracks of the multi-track disposition and the individuation of systems. In the case of the ideal track, for instance, we are just considering the falling stone on its own without any interfering factor, while in the case of the other tracks we are considering more inclusive systems, namely, the stone plus a medium. In the case of such an inclusive system, the multi-track disposition that we have considered so far is a disposition of one part of the system only, namely, a disposition of the stone. However, that is not the only disposition to consider. We can furthermore attribute a disposition to the compound or inclusive system (including, for instance, the medium) to behave in a certain way. (We have seen in the previous chapter that, at least in physics, there are explicit laws of composition that determine how the behaviour of a compound system is disposed given what the subsystems are disposed to.)

One thing that should be noted is that even though the individuation of a system fixes the track, e.g., the ideal track, this on its own does not yet determine the quasi-inertial behaviour completely because the exact quasi-inertial behaviour furthermore depends on initial conditions. Thus, for example, Newton's first law does not determine the quasi-inertial behaviour of isolated bodies completely. Complete determination requires the relevant initial conditions. Thus, Newton's first law tells us that in the absence of impressed force the velocity and direction of a body does not change. It does not give us the actual direction and velocity of a body.

What all this amounts to is the following: Even though it is true that systems may be disposed to many different kinds of behaviour depending on the presence or absence of interfering factors, it is also true that once we have identified the inclusive system we are interested in, e.g., the falling stone in water, and once the initial conditions are fixed, there is only one kind of behaviour this inclusive system is disposed to *provided there are no interfering factors*. So, once we have identified what system we are interested in and we focus on the ideal track (no (further) interfering factors), and given the initial conditions, we have narrowed down the options the system has to exactly one kind of behaviour. It is this behaviour that I have referred to as the 'quasi-inertial' behaviour of a system. As long as we are not considering indeterministic settings, once the system is specified there is only one kind of behaviour it is disposed to, and this disposition is an intrinsic feature of the (compound) system under consideration.

The identification of quasi-inertial processes is, as we have seen, relative to a prior identification or specification of the systems. I might, for instance, be interested in falling objects in a vacuum. Once this has been settled, it is – as already mentioned – an objective matter what the quasi-inertial behaviour is. In other circumstances, however, I might be interested in the behaviour of falling objects in a medium; then the system's quasi-inertial behaviour will be different. But that is not surprising, given that we are looking at different systems. Similarly, we might want to examine the quasi-inertial temporal development of a system consisting of prey and predator populations, say, foxes and rabbits (in a certain environment). The behaviour of such a system can be described by the Lotka–Volterra equations. However, there might be another situation in which we simply want to know the quasi-inertial temporal development of the population of foxes (in a given environment). Biology does not tell us what kind of system we should study; instead, it gives us information about the quasi-inertial behaviour of the systems that we have chosen.

This point helps to address a worry that has been raised by Blanchard and Schaffer. In the context of constructing causal models with default values, they observe that

> default status seems conflictingly overdetermined in many cases. Suppose [. . .] that Sally (like most other drivers on the road with her) is driving 65 mph in a 55 mph zone, and gets into an accident that would not have occurred had Sally been driving at 55 mph. [. . .] But how are we supposed to assign default or deviant values to [the variable] Speed? The social norm is to

drive 55 mph, but the statistical norm is to drive 65 mph. (Blanchard and Schaffer 2017, 193)

What I call 'quasi-inertial' behaviour is relevantly different in two respects from the notion of 'default' behaviour considered by Schaffer and Blanchard. First, they consider systems (drivers) with different options (55 mph, 65 mph, etc.). By contrast, the quasi-inertial behaviour of systems we were discussing is completely determined once the initial conditions are fixed. So in the case of Sally, either the system is deterministic but we do not know the initial conditions of the system, and thus, even though it is determined, we simply don't know what the quasi-inertial behaviour is; or the system is indeterministic and thus falling outside the scope of quasi-inertial behaviour considered in this chapter so far (see Section 3.8 for dealing with indeterministic quasi-inertial processes). Second, for Schaffer and Blanchard it is *external* considerations (relative to the system, i.e., relative to a single driver such as Sally) that determine whether one of these options, i.e., one of these behaviours, can be classified as 'default' (social norms, statistics). Neither point applies to the quasi-inertial behaviour I have discussed. There is exactly one way in which the free-falling stone can fall in a vacuum. Its quasi-inertial behaviour is due to intrinsic features of the system and completely independent of system-external social norms or statistical considerations.

So far, I have argued that Schaffer and Blanchard's considerations do not apply to falling stones and other systems I have discussed. But what about Sally? What is her quasi-inertial behaviour when sitting in a car on the highway? Let me start by discussing a slightly less complex case.

Take the rabbit: Is its quasi-inertial behaviour to continue to live or to die? This is an ambiguous case. Note, however, that the ambiguity is entirely due to the fact that the question does not uniquely specify what the relevant system is. If we are considering the rabbit completely on its own, i.e., in a world without oxygen, grass, etc., its quasi-inertial behaviour presumably is to die pretty soon. If, on the other hand, the question is meant to refer to a rabbit in a different context, i.e., to a different system, the answer may very well be different as well. In other words: The quasi-inertial behaviour of the system in question depends on features of the system, and the question of the quasi-inertial behaviour of 'the rabbit' is too vague to allow a unique answer. The same reasoning applies to Sally. Absent further specifications of the system *Sally on the highway*, the best we can get at empirically will presumably be a probability distribution over different speeds. It is only if we add details to the system, e.g., whether and under what circumstances Sally complies with the social norm, that we will get a definite quasi-inertial behaviour.

Given the multiplicity of systems we might individuate and the corresponding multiplicity of quasi-inertial processes, one might wonder about the implications for causal explanation. In a causal explanation we typically (though not always) attribute *a* cause rather than a multiplicity of causes. What is the mechanism that slashes the multitude of quasi-inertial processes? The essential point is that in a causal explanation we start with the specification of an effect. Something has happened, often something unexpected. In a causal explanation, what has happened is contrasted with another course of events, another process, that would have led to the non-occurrence of the effect. By this procedure, the first two conditions for causation discussed in Section 3.4 (as applying to a window that is shattered by a stone) are specified:

(1) An event *e* (the effect) obtains: System S (the window) is in state Z_e at t (it is shattered).
(2) The quasi-inertial behaviour of S would have led to S being intact at t (S in state Z_i, which is different from Z_e): In the absence of interfering factors the window, which as a matter of fact is broken, would have remained intact.

Thus, it is by specifying the effect that the multitude of possible systems/ inertial processes (in this case, for example, (i) the window pane remaining intact or (ii) the window pane and the stone developing into a shattered window and a stone lying around somewhere) are reduced to just one (in this case, the window pane remaining intact) by contrasting them with the effect in question.

Even though in our ordinary practice of causal explanation these specifications are fairly vague, the important point is that the specification of the effect fixes (to some extent) the relevant quasi-inertial behaviour (as well as the system involved).

3.5.3 Interferences

What goes for quasi-inertial processes goes for disturbing or interfering factors as well. The case of Newtonian physics is particularly simple: The first law not only explicitly specifies a system's (quasi-)inertial behaviour; it furthermore states what the possible interfering factors are ('impressed forces'). Moreover, the second law describes the exact influences of these interfering factors (it is a 'law of deviation'; Maudlin 2004, 431). Newton's laws thus give us two kinds of information that allow us to characterise what the relevant interfering factors are: First, they tell us what candidate interfering factors there are ('impressed forces'), and second, they tell us exactly how these factors, if interacting, modify the quasi-inertial behaviour.

In other disciplines, there are no general and explicit accounts of interfering factors, but it seems plausible to argue that there is implicit knowledge concerning what counts as a legitimate interfering factor and how such a factor might qualitatively modify the envisaged quasi-inertial behaviour. In economics, for example, state interventions or the decisions of the federal reserve bank might be considered as candidates for interfering factors, and we also know (or at least form hypotheses about) how, for instance, the change of certain interest rates modifies what can be considered the quasi-inertial behaviour of an economic system. Both quasi-inertial processes and interfering factors are identified in the sciences on a case-by-case basis.

What this account provides is thus a rather *thin* concept of actual causation: A description of what happens in terms of 'cause' and 'effect' is a rather abstract and coarse-grained description. A more fine-grained description will turn to the details that are provided by the underlying physics, biology or economics. However, there is something that these cases of actual causation have in common and in virtue of which causal claims are true: First, there is a system that is disposed to a specific quasi-inertial behaviour, and second, the quasi-inertial process is not displayed due to an interfering factor.

3.5.4 Worry

Let me mention a worry. Isn't the account circular in the sense that it spells out causal terminology in terms of quasi-inertial processes and interfering factors, which in turn are causally laden terms? Another way of putting this worry is: Doesn't the account fail to be reductive? My response is that the account presented here *is* a reductive account, at least in one sense of 'reduction'. Accounts of causation can be reductive in various ways. First, one might attempt to define or explicate the concept of causation without relying on causal terminology ('semantic reductionism'). Second, one might try to provide an account of how causal claims are tested without relying on prior knowledge of (other) causes ('epistemological reductionism'). I do not claim that the account presented here is reductive in either the semantic or the epistemological sense. Third, however, one might try to give a reductive ontological account of causation. This last one comes in (at least) two varieties. The first attempts to reduce causal facts to noncausal facts, while the second attempts to reduce causal facts to facts described by underlying sciences, i.e., to the underlying dynamic laws plus certain initial and other conditions. These two versions of ontological reduction may not coincide: if basic physical assumptions of fundamental physics turn out to be causal assumptions, as Frisch argues (Frisch 2014), a reduction to physical facts would not amount to a reduction to non-causal facts. My focus is not the reduction of causal facts to non-causal facts but rather, the reduction of macroscopic causal relations to facts described by law statements or generalisations of the underlying sciences, such as physics or biology (whether causal or not). I should also note that a pragmatic or perspectival element comes into play via the specification of the effect. So, the reductive project is not overly ambitious: the claim is that once the effect is specified, the causal facts that account for the effect can be characterised in terms of generalisations of the underlying sciences. Thus, the metaphysical assumptions introduced in Chapter 1 and Chapter 2 suffice to account for our practice of causal explanation.

3.6   Relation to Other Process Theories and to Dispositional Accounts of Causation

3.6.1   Process Theories of Causation

I argued that the application of token-causal terminology can be best understood if a cause is taken to be an interfering factor to the quasi-inertial behaviour that a system is disposed to display. This is a claim about processes, because the quasi-inertial behaviour concerns the temporal evolution of a system (e.g., the process that is described in Newton's first law). So, how exactly is this view related to traditional process theories?

Process theories tend to take 'causation to be the transfer or persistence of properties of a specific sort' (Dowe 2009, 214). The quasi-inertial processes we have talked about can indeed be characterised in terms of the persistence of properties: The behaviour in question is persistently manifest as long as nothing intervenes. Process theories consider interferences with these processes as cases of causation. Thus far I agree. There are, however, two important points of disagreement: First, according to the definition (DC) of the disruptive concept of causation, the processes themselves do not constitute cases of token causation. A statue being at a certain place at 2 p.m. today is not a cause of its being there at 5 p.m., according to this definition. Therefore, I do not talk about *causal* processes but rather about *quasi-inertial* processes. Second, the characterisation of the relevant processes and disturbances (interactions) is different. Whereas Dowe and Salmon define these processes either by the mark criterion or in terms of invariant or conserved quantities (Dowe 2009 provides an overview), I characterise them in terms of the underlying systems' dispositions. A ball rolling on a flat surface is described by traditional process theorists as a causal process because it conserves kinetic energy and momentum. I characterise this process as a quasi-inertial process because it manifests a disposition, namely, the one that is attributed to it in Newton's first law. To qualify as a quasi-inertial process, it is not necessary to have certain invariant/conserved physical properties. The essential question is whether or not the relevant system has a certain disposition, and determining this is the responsibility of the sciences: It is for the physicists to decide whether or not bodies have the disposition to continue in uniform rectilinear motion if no forces are impressed on them. Similarly, it is for economists to decide whether or not an economic system in which inflation rises will yield higher unemployment rates (provided nothing interferes and certain factors are held constant).

A consequence of this account is that whether or not something qualifies as a quasi-inertial process need not be spelled out in terms of physics. To the extent that biology or economics attributes dispositions to systems that concern their temporal evolution, these disciplines deal with biological or economical quasi-inertial processes.

Similarly, what qualifies as an interference with a quasi-inertial process need not be spelled out in terms of physics. I want to leave room for various kinds of interferences that have to be specified locally. It is the job of the sciences in question to provide a more detailed description of the interference. While physics considers a certain repertoire of interfering factors that can be described in terms of physical interaction or, maybe, conserved quantities, economics considers state interventions, decisions of the federal reserve bank or natural catastrophes.

The account of causation in terms of quasi-inertial processes and inferring factors is able to deal with all three issues that Dowe (2009) has identified as major problems for process theories. The first one is the problem of reduction: Because the account presented here allows for causal relations obtaining at various 'levels', it is not committed to the claim that all causation is ultimately physical causation. The second issue is the problem of negative causation, and the third issue concerns what has been called the 'problem of misconnection': Both will be dealt with in Chapter 4.

3.6.2 Dispositional Accounts of Causation

I explicated causation as an interference with a quasi-inertial process that a system is disposed to display. That raises the question of how the account presented here is related to other dispositional accounts. Very briefly, there are various ways in which dispositions might be relevant for various accounts of causation if one assumes that law statements attribute dispositions to systems:

(i)   It might be argued that dispositions that are attributed in law statements give rise to regularities, and that causation needs to be explicated in terms of these regularities. Such an account would inherit the problems the regularity theory is confronted with.

(ii) Similarly, it could be argued that dispositions are the truth-makers for certain counterfactual claims in terms of which the cause–effect relation needs to be explicated. In an earlier publication (Hüttemann 2004) I briefly sketched this option. Such an account, however, inherits the problems of the counterfactual theory of causation.

(iii) According to the major or best-known dispositional account of causation, a cause is simply a disposition manifesting itself (Mumford and Anjum 2011). This characterisation of causes follows from the definition of dispositions if the latter are defined as properties that are oriented towards certain *causal* effects. There is not enough space to discuss the pros and cons of this account (see McKitrick et al. 2013 and Chakravartty 2013 for discussion). I just want to point to two differences between this account and the one presented earlier. First, while according to the disposition-as-causes account, a cause is a disposition that manifests itself, what is essential to the account presented earlier is a disposition whose manifestation is interfered with (see Mumford 2014, 337). Second, because I do not build causation into the definition of dispositions, my account of causation allows the position that causation is essentially a macroscopic feature that need not be implemented at the micro-level, even if there are dispositions at the micro-level.

3.7 The Function of the Disruptive Concept of Causation and the Role of Causal Intuitions

Once the disruptive concept of causation is spelled out in terms of interferences and quasi-inertial processes, the role causation plays in certain everyday contexts can be understood. Why is it that causation plays such a significant role in explanation, in the attribution of responsibility, and for purposes of manipulation? In causally explaining an event ('Why did the building collapse?'), very often we are interested in what interfered with a quasi-inertial process that would have led to the effect not occurring. And that is exactly what the disruptive concept of causation picks out. The same holds for the role of causation in the context of attributing responsibility. What we are interested in is whether a human action interfered with a process that would have led to the effect not occurring had the interference not taken place. Furthermore, causal knowledge in the sense outlined here helps us to interfere with processes in ways that may be useful for us.

3.8 The Focal Concept of Causation and Other Concepts of Causation

So far, I have provided the conceptual groundwork for the disruptive concept of causation. As I indicated at the outset, such an explication does not yield a comprehensive theory of causation. More specifically, there are other concepts of causation that play a role in the sciences. The concept explicated so far is, however, important for understanding these other uses of 'cause' as well. In this section I will illustrate how further concepts of cause are related to the focal concept. (So, these conceptions are not competing but rather, complementary conceptions.)

3.8.1 *Closed-System Causation*

We have already seen how Newton's first and second laws of motion can be easily understood as incorporating causation in terms of quasi-inertial processes and interferences. During the nineteenth century, however, another notion of causation emerged in the context of physics. This is evident, for instance, in Laplace's well-known characterisation of determinism: 'We may regard the present state of the universe as the effect of its past and the cause of its future' (Laplace, *A Philosophical Essay on Probabilities*, quoted in Earman 1986, 7). In this context, a cause is not something that interferes with a quasi-inertial process but rather, an (earlier) state of a system that determines another (later) state of the same system. Interestingly, Mach noticed this development early on:

> The present tendency of physics is to represent every phenomenon as a function of other phenomena and of certain spatial and temporal positions. [. . .]. *Thus the law of causality is sufficiently characterised by saying that it is the presupposition of the mutual dependence of phenomena.* Certain idle questions, for example, whether the cause precedes or is simultaneous with the effect, then vanish by themselves. The law of causality is identical with the supposition that between the natural phenomena $\alpha, \beta, \gamma, \delta, \ldots \omega$ certain equations subsist. The law of causality says nothing about the number or form of these equations; it is the problem of positive natural investigation to determine this. (Mach 1872, 61; original emphasis)

While in his later writings Mach uses this observation to argue that the concept of causation has no place in science and should be *replaced* by the concept of function, in his early writings he suggests that in the light of the development in the sciences we should *redefine* our notion of cause:

> Let us call the totality of the phenomena on which a phenomenon a can be considered as dependent, *the cause*. If this totality is given, a is determined, and determined uniquely. Thus the law of causality may also be expressed in the form: 'The effect is determined by the cause.' (Mach 1872, 63–4)

Mach reacts to a development that has taken place during the nineteenth century: the description of closed systems in terms of differential equations.[5] This development led to a new conception of what causation might mean in physics. This emergence of a new concept of causation was also noted later on, e.g., by the theoretical physicist Peter Havas, who explicitly distinguished the ordinary concept of cause (the disruptive concept of causation) from the use of causal terminology by physicists:

> We are all familiar with the everyday usage of the words 'cause' and 'effect'; it frequently implies the interference by an outside agent (whether human or not), the 'cause', with a system, which then experiences the 'effect' of this interference. When we talk of the principle of causality in physics, however, we usually do not think of specific cause-effect relations or of deliberate intervention in a system but in terms of theories, which allow (at least in principle) the calculation of the future state of the system under consideration from data specified at time t_0. No specific reference to 'cause' or 'effects' is needed, customary, or useful, but it is understood that all the phenomena (or variables) which can influence the system have

---

[5] See Scheibe (2006, 223–6) for a discussion of this development; I follow his account.

been taken into account in the initial specification, i.e., that the system is closed. (Havas 1974, 24)

According to closed-system causation, one state of a (closed) system causes another state of the same system to obtain iff the former (uniquely) determines the latter (by virtue of the laws of nature).[6] When in the context of quantum mechanics the so-called 'principle of causality' came under attack, it was this concept of causation ('closed-system causation') that was under discussion rather than the disruptive concept of causation.

The most important difference between these concepts is highlighted by the fact that causation in terms of interferences in quasi-inertial processes disrupts exactly what closed-system causation presupposes: that the system in question is closed (see Scheibe 2006, 223). However, there are also important commonalities, which explain why the term 'cause' is used in both cases. In both cases, quasi-inertial processes are central, and in both cases, the cause *determines* the effect. Given the laws of nature, in the case of interference this determination is conditional on the quasi-inertial process, whereas in the case of closed-system causation an earlier state of a quasi-inertial process completely or uniquely determines a later state (provided the laws are deterministic).

A further connection between these two concepts of causation is the following: When we are dealing with a disruptive cause, i.e., with an interfering factor in a quasi-inertial process, the interfering factor plus the system with which it has interfered can be considered as a compound closed system. Thus, what is going on when we have a case of disruptive causation can also be described in terms of closed-system causation. Conversely, if in a closed system we can identify subsystems, the behaviour of one of these subsystems may sometimes be characterised as caused by the interference of another subsystem.

While closed-system causation may be important in understanding how the expression 'cause' is used in some of the sciences, notably in physics, it does not seem to be what is relevant in the social sciences or what is tracked by our causal intuitions. We don't usually want to say that the event of my chair standing in front of the table at 9am causes the chair to stand in the same place 5 minutes later.

Be that as it may, a comprehensive theory of causation has to take into account causal concepts, whether or not they are tracked by our

---

[6] Another theoretical physicist, Friedrich Hund, likewise argued for these two concepts of causation and observed with respect to what I called 'closed-system causation' that – in contrast to our ordinary concept of causation – it does not entail temporal asymmetry (quoted in Scheibe 2006, 225).

intuitions. This is important for understanding our causal discourse, which comprises a number of different concepts. Closed system causation, for instance, is relevant for a more comprehensive understanding of token causation. While the account of causation in terms of interferences with quasi-inertial processes allows us to say that an event has *more than one* cause, it does not cover what Mill called 'total cause' (Section 3.2). Mill's total cause should be understood as a complete state of a system and thus as a cause in the sense of closed-system causation.

3.8.2 Probabilistic Causation

There is a tradition that tries to analyse all cases of causation in terms of probabilistic concepts. Even if this program is too ambitious, it is certainly true that there are cases in which probabilistic relations are taken to be evidence for causal relations or are perhaps even taken to constitute causal relations. In many sciences, causal reasoning is probabilistic. The basic idea in this tradition is that causes raise the probability of the occurrences of effects. As is well known, there are a number of problems with this idea (for a discussion see Hitchcock 2016). I will not go into the details and qualifications that are needed to make probabilistic causation a viable concept.

Let me explain how probabilistic causation is related to the conception of disruptive causation developed earlier. Probabilistic causal claims such as 'Smoking causes lung cancer' are typically type-level causal claims, in contrast to the token-level claims I have considered up until now. There are two different ways in which the truth of probabilistic causal type-level claims may depend on underlying processes and interferences, depending on whether the underlying processes are deterministic or genuinely indeterministic.

(i) Deterministic case: The probabilistic type-level claim obtains in virtue of coarse graining over different kinds of causing. Suppose the claim is that some event type X probabilistically causes Y to occur. This may be true in virtue of the fact that two different kinds of situations are involved:

Situation type A: X disturbs process P, Y occurs as a result of ensuing process P*.

Situation type B: X disturbs process P, X* disturbs ensuing process P*, Y does not occur.

Coarse graining over heterogeneous situations of type A and type B makes probabilistic type-level claims true. If $Pr(Y \mid X) > Pr(Y)$ holds (and some further conditions such as screening off), one might say that X causes Y. In this deterministic case, no new element needs to be introduced in order to understand probabilistic causal claims at the type level.

The situation is different if one wants to accept genuine chancy causation:

(ii)   Indeterministic case: The probabilistic type-level claim obtains in virtue of genuine chancy causation. Suppose the chance of receiving a certain illness I is 0.5 percent. Excessive consumption of X raises the probability to 2 percent. We furthermore assume that the relevant type-level claim 'Excessive consumption of X *causes* I' cannot be explained in terms of coarse graining over heterogeneous situations. This case requires the introduction of genuinely indeterministic processes, i.e., processes that have various outcomes and a probability distribution over the outcomes. It might be the case that human beings (in the usual contexts) develop illness I in 0.5 percent of the cases and fail to develop I (i.e., non-I) in 99.5 percent of the cases. Take this to be the quasi-inertial process. I take it that x is the cause of i (lower case for instantiations of types) if the person in question excessively consumes x, if the person furthermore develops I, and if the original probability distribution changes in virtue of the consumption of x such that it is more likely the person develops I.

Genuine chancy causation (at the token level) can then be understood as follows:

x causes i, if x occurs, i occurs, and the quasi-inertial process probability distribution has been disturbed by x so that $Pr(I \mid X) > Pr(I)$ is true.

The central idea is that genuinely indeterministic processes have a probability distribution over outcomes, which might be disturbed by interfering factors. If the disturbance raises the probability of the outcome, then the interfering factor may be said to have caused a certain outcome. This is an extension of the original account to the case of chancy causation. (It might also be extended to the coarse-graining case discussed earlier.)

3.9   Mach and Russell Revisited

Let me briefly return to Mach's and Russell's worries concerning the concept of causation, which I discussed at the outset of the chapter. One issue they were concerned with was whether our ordinary concept of causation is compatible with the way physics describes the world. The first thing to note is that some of their arguments presuppose that our ordinary concept of causation ought to be spelled out in terms of a regularity view of causation (see Section 3.2). These arguments, however, fail to convince because, as we have seen, our ordinary concept of causation should be explicated in terms of a system's quasi-inertial behaviour and an interfering factor. Regularities are not required for causation to obtain.

On the other hand, Mach and Russell are completely right in observing that the disruptive concept of causation is not what physicists focus on when they describe the behaviour of systems in terms of differential equations. To repeat what the theoretical physicist Havas observed:

> When we talk of the principle of causality in physics, . . . [we think] in terms of theories, which allow (at least in principle) the calculation of the future state of the system under consideration from data specified at time t_0. No specific reference to 'cause' or 'effects' is needed, customary, or useful, but it is understood that all the phenomena (or variables), which can influence the system have been taken into account in the initial specification, i.e., that the system is *closed*. (Havas 1974, 24)

Whether or not we are willing to call the state at one time the cause of the state at another time is ultimately merely a matter of labelling. The essential point is that Mach and Russell saw that differential equations describe mutual determination, which is different from our ordinary (disruptive) concept of causation.

However, a stronger claim that Mach and Russell probably defended is the view that what our ordinary concept of causation requires is incompatible with the way physics describes the world. This latter claim needs to be rejected. Russell's main argument was that causation in this sense is time-asymmetric and there is nothing in the fundamental equations that allows for this. However, asymmetric processes at the macro-level are compatible with the time-asymmetry of the fundamental equations.[7] Thus, even though the exact relation between macro-level and micro-level laws is a controversial issue, it is generally assumed that the time-asymmetric character of the second law of thermodynamics is compatible with the fundamental equations of motion being time-symmetric.

The following picture emerges: Even if we had a complete description of the development of the universe in terms of differential equations, there would still be room for our ordinary (disruptive) concept of causation provided the following conditions were met. First, within the universe we

---

[7] How exactly macroscopic temporal asymmetries are related to the underlying physics is a matter of debate. See, e.g., Albert 2000, Albert 2015, Kutach 2011 or Zeh 2001.

are able to identify subsystems to which we can attribute quasi-inertial behaviour. Second, the quasi-inertial processes are either time-asymmetric or – as is sometimes argued – embedded in contexts in which the majority of processes are time-asymmetric. The application of our ordinary concept of causation may be restricted to contexts in which the conditions are met, e.g., to the macroscopic realm.

3.10 Metaphysical Conclusion

I have argued that the disruptive concept of cause can be spelled out in terms of interferences and quasi-inertial processes, i.e., processes that systems are disposed to provided there are no interfering factors. No additional assumptions over and above those introduced in Chapters 1 and 2 are needed to account for our practice of causal explanation.

These interferences and quasi-inertial processes can in every single case be explicated further in scientific terms, i.e., ultimately in terms of scientific generalisations concerning dispositions of systems. In other words, I have shown how what we pick out by the disruptive concept of cause is integrated into a world as described by the sciences in terms of laws of nature and assumptions about prevailing conditions. This implies in particular that causal dependence (i.e., the modal dependence of the effect on the cause) can be accounted for in terms of nomological dependence and, thus, ultimately in terms of invariance relations.

The disruptive concept of causation is important for at least two reasons. First, it is the conceptual basis for other causal concepts such as probabilistic causation or the structural equation concept (see next chapter), concepts that play an important role, e.g., in the social sciences. Second, as I will argue in more detail in the next chapter, it is the concept that is tracked by many of our causal intuitions.

With respect to closed-system causation, I have shown how it is related to the disruptive concept of causation and what developments in the sciences gave rise to it. For this concept to be applicable to systems or phenomena in the world – just as in the case of the disruptive concept – nothing has to be assumed beyond the obtaining of scientific facts, i.e., facts concerning laws of nature. There is nothing in our practice of working with causal concepts that requires any additional metaphysical assumptions over and above those discussed in Chapters 1 and 2.

CHAPTER 4

Causation – Application and Augmentation

In this chapter, I will put to work the account of causation in terms of quasi-inertial processes and interferences, which I outlined in Chapter 3. More particularly, I will examine how the account fares in dealing with a number of prominent problems that have been discussed in the causation literature, e.g., the pre-emption problem (Section 4.1) and the issue of the transitivity of causation (Section 4.4). It will turn out that our original definition of disruptive causation needs to be augmented. Discussing these issues will help me to address two objections that process theories of causation traditionally have to face, the problem of misconnection (in Section 4.1) and the problem of disconnection (Section 4.2). I will close by briefly comparing the account outlined here with the structural equations approach.

In this chapter, the appeal to causal intuitions will play a significant role. This may give rise to some worries. Dowe made a twofold distinction between a conceptual analysis of causation, on the one hand, and an empirical analysis, on the other. The conceptual analysis of causal concepts aims to explicate the concepts through an appeal to causal intuitions, while an empirical analysis 'seeks to establish what causation in fact *is* in the actual world' (Dowe 2000, 3). Given that my ontological project falls into the latter category (and given the scepticism I voiced in the Introduction concerning the appeal to intuitions when it comes to metaphysical issues), one might wonder whether my reliance on causal intuitions and causal judgements of ordinary, competent English speakers might be misplaced.

While I generally share the scepticism regarding the relevance of conceptual analyses for metaphysical claims, I take the situation to be special in the case of causal concepts. Causal reasoning, causal beliefs and causal concepts are deeply engrained in our cognition and may be arguably at least in part neurologically hard-wired (see Danks 2009 for an overview). From an evolutionary perspective it makes sense to assume that our causal intuitions by and large carve nature at its joints. Thus, in the special case of causal intuitions we have good reasons to consider these intuitions as evidence for features of the actual world. Furthermore, it makes sense to assume that these intuitions play a role in causal explanation not only in everyday contexts but also in contexts of scientific causal explanation, i.e., in causal explanation that relies on scientifically established theories or laws. Even though I have to admit that I know of no empirical study that discusses or substantiates my assumption that causal reasoning in scientific contexts is shaped by causal intuitions, I take it to be fairly plausible. For instance, when engineers try to establish the cause(s) of the collapse of a particular bridge and take into consideration a recent flood, the long-term consequences of an earthquake as well as poor maintenance, it seems plausible that causal intuitions concerning pre-emption and other forms of overdetermination might play a role. I thus take it that with respect to *causal* intuitions two conditions are met, namely, (1) that they shape our practice of causal reasoning and (2) that they are truth conducive. Because these two conditions are met it makes sense to turn to these intuitions within an account of the metaphysics of scientific practice.[1]

My aim is to show that these causal intuitions can be reconstructed as intuitions about quasi-inertial behaviour and interfering factors.

4.1   Pre-emption

In Chapter 3, I have elaborated a suggestion concerning which scientific facts we pick out when we use causal terminology in ordinary contexts, namely, facts about interfering factors with quasi-inertial processes. In this section I will argue that such a concept of cause accounts for at least the causal intuitions we have in pre-emption scenarios.

Consider the case of late pre-emption. Suzy and Billy both throw stones at a window; Suzy's stone gets there first and shatters the window. Billy's stone arrives at the scene a bit later but does not destroy the window because the latter is already shattered.

As is well known (Lewis 1986; Collins, Hall and Paul 2004, 22–3), late pre-emption is a problem for (simple) counterfactual accounts of causation: It is not true that if Suzy had not thrown the stone, the window would have remained intact.

If the account in terms of quasi-inertial processes and interfering factors is correct, we have an explanation of our intuitions in the case of late pre-emption: The effect in question is the shattering of the windowpane at t

---

[1] Thanks to an anonymous referee for raising this issue.

(the windowpane is our system S; Z_e is its state of being shattered). According to our definition (see Section 3.4), when we consider an event c to be a cause of another event e, we assume that c is the event of some factor interfering with a system S such that S develops into a state Z_e at t, rather than some state Z_i at t into which S would have developed if the quasi-inertial process had been displayed. The effect we want to account for is the windowpane being shattered rather than non-shattered. Thus, the windowpane sitting in its frame and remaining intact is the system's quasi-inertial behaviour in question. Without an interference the quasi-inertial process would have resulted in the window not being shattered at t (state Z_i). As a matter of fact, the windowpane's quasi-inertial behaviour has been interfered with by Suzy's stone, such that the quasi-inertial behaviour has not been displayed. Thus, according to our account it was Suzy's stone that caused the shattering of the window.[2] By contrast, the windowpane's quasi-inertial behaviour has not been interfered with by Billy's stone, such that the quasi-inertial behaviour has not been displayed. There was, e.g., no energy transfer from Billy's stone to the windowpane, because the windowpane was all already shattered when Billy's stone arrived. So, Billy's stone did not cause the shattering of the window.

The account just outlined is what one would expect from a process theoretical perspective on causation. Causation does not consist in counterfactual dependence but rather in the fact that a process of a certain specified kind has been interfered with. However, there are problems with the process theoretical solution to the pre-emption problem. The main point is that relevant interferences have to be distinguished from irrelevant interferences. This has been dubbed the 'problem of misconnection' (Dowe 2009, 221). Let me approach this issue by discussing a number of objections to the foregoing solution of the pre-emption problem.

4.1.1   Objections

First, the case is pretty straightforward as long as we assume gravity to be switched off. However, if we switch on gravity we can ask: What about Billy's stone – didn't it interfere with the windowpane too, by virtue of a gravitational interaction? Clearly the *manner* in which the window shattered would have been different had Billy's stone not been present at

---

[2] Strictly speaking, the stone hitting the pane is the cause of the shattering. Suzy throwing the stone is the cause of the shattering only if we allow it to be the case that if c causes d and d causes e, c is a cause of e (maybe provided further conditions obtain – the issue of the transitivity of causation is dealt with in Section 4.4).

all. Thus, we should count Billy's throwing of the stone as an interfering factor – rather than a cause and a pre-empted cause, we would have two contributing causes (for a discussion see Paul and Hall 2013, 56–7). But that would not capture our intuitions in this case, according to which Suzy's throwing of a stone is the only cause.

By way of rejoinder, let me start by specifying what the process theorist is committed to. A process theorist argues that causation is a matter of interference with certain kinds of processes (in my case, quasi-inertial processes). If there are various interferences, the question is whether these interferences were relevant for bringing about the effect. Thus, in order to account for our intuitions in the modified pre-emption case it does not suffice to merely identify the interferences; we furthermore have to make a distinction between those interferences that are relevant and those that are not. (Those that are *relevant* are interferences *such that* the effect occurs; see definition (DC) in Section 3.4). Being an interference is thus not sufficient for being a cause. A cause is a *relevant* interference.

So, why is the gravitational interaction of Billy's stone not relevant for the shattering of the window? An essential point is that when we want to specify a cause of a certain event, we first have to specify the effect for which we seek the cause. Remember that our notion of disruptive causation is contrastive with respect to the effect. It makes a difference whether we are looking for the cause of there being a noise (rather than no noise) or whether we are asking for the cause of there being a loud noise (rather than a moderate noise). Depending on how we specify the effect, we need to consider different systems/quasi-inertial processes and different interfering factors. So, if the effect is the shattering of the window (in contrast to its remaining intact), Suzy's stone striking the window will be the *relevant* interfering factor. The gravitational interaction that has also taken place between Billy's stone and the windowpane is too small to cause the window to shatter. We know this in virtue of physical laws of interaction, conservation of energy, etc. If there had only been the gravitational interaction with Billy's stone, the window would have remained intact. Conversely, if there had only been Suzy's throw, the shattering would have occurred anyway. Only Suzy's stone interacts with the system S (the window) *such that* S at t is in state Z_e rather than in state Z_i and thus qualifies as an interfering factor for the effect in question. I suggest therefore that there is a counterfactual test that allows us to isolate the *relevant* interacting or interfering factors. The relevant interacting factors are those that interact *such that* the effect occurs.

Second, I used counterfactuals to help identify which throwing inter-feres with the quasi-inertial process such that the effect is brought about. But what counterfactuals are we to consider, and why are we to consider those specific counterfactuals as opposed to some others? The following counterfactuals seem plausibly relevant to the case at hand: (a) if Suzy had not thrown her stone *in the absence of Billy's stone*, then the shattering of the window would not have occurred and (b) if Billy had not thrown his *in the absence of Suzy's stone*, then shattering of the window would not have occurred. Both (a) and (b) are true. Why are these counterfactuals not the ones relevant to determining which event interferes with the quasi-inertial process of the window remaining intact?

The answer is that the counterfactual test has a particular purpose. The purpose of the test is to find out which of the various *actual* interferences bring about the effect, i.e., which interferences are those that are relevant for the occurrence of the effect (given all the other interferences) and which are irrelevant for the occurrence of the effect (given all the other interferences). So, the recipe for determining whether an interference is relevant for a particular cause (or *such that* the effect occurs) consists in considering the following counterfactual: In the antecedent we consider a situation in which all the interferences are as they are in the actual situation, with the exception of the inference with respect to which we are trying to figure out whether it is relevant. The interference whose relevance we consider needs to be eliminated. If, despite the elimination, the effect occurs, the interference is irrelevant. If, given the elimination, the effect fails to obtain, the interference in question is relevant.

Thus, the following counterfactuals need to be considered to determine whether a particular interference into a quasi-inertial process was relevant for the shattering of the window:

(A)   If Suzy had not thrown the stone (and thus the interference in terms of momentum and energy transfer from Suzy's stone had not taken place), the window would not have been shattered, provided (i) all the other interferences/interactions (except those interferences that occur as a consequence of the interaction under consideration, i.e. Suzy's throw)[3] are held fixed – and that includes in particular that

---

[3] Why is there this exception-clause? Suzy's throw is not the immediate cause of the shattering. This gives rise to the following consideration: We have at least two interferences: (i) Suzy with the stone, and (ii) the stone with the window. Then, however, it is not true that 'If Suzy had not thrown the stone (and thus the interference in terms of momentum and energy transfer from Suzy's stone had not taken place), the window would not have been shattered, provided all the other interferences are held fixed.' For, holding the other interferences fixed (other than Suzy's transference of momentum) implies

Billy's stone interacts with the window only via gravitation and not via momentum or energy transfer, and (ii) no new interferences have been added.

This counterfactual comes out as true; therefore, the throwing of the stone by Suzy (or, rather, the interference of the stone in terms of momentum and energy transfer) is a relevant interference for the shattering of the window.

(B) If Billy had not thrown the stone (and thus the interference in terms of gravity had not taken place or had been weaker), the window would not have been shattered, provided (i) all the other interferences/ interactions (except those interferences that occur as a consequence of the interaction under consideration, i.e., the gravity due to Billy's throw) are held fixed – and that includes in particular that Suzy's stone interacts with the window only via momentum or energy transfer, and (ii) no new interferences have been added.

This counterfactual comes out as false; therefore, the throwing of the stone by Billy (or rather the interference of the stone in terms of gravity) is an irrelevant interference for the shattering of the window.

For the purpose of distinguishing relevant from irrelevant interferences *among those interferences that actually occur* the counterfactuals to consider are (A) and (B), not (a) and (b).

To sum up the considerations concerning relevance:

> An interference into a quasi-inertial process is *relevant* with respect to an effect *e* (it is an interference *such that* the effect *e* occurs) iff had the interference not occurred, the effect would not have occurred either – provided (i) all the other interferences (except those interferences that occur as a consequence of the interaction under consideration) are held fixed and (ii) no new interferences have been added.[4]

Third, counterfactual reasoning plays an important role when it comes to the question of whether an interacting factor is a relevant interfering factor for a certain effect. Does this turn our analysis into a counterfactual

---

holding fixed that the stone interferes with the window. So, the window still breaks. The clause in brackets helps to avoid this problem. (Thanks to an anonymous referee for raising this point.)

[4] My criterion for 'relevance', which I introduced to solve a problem that traditional process-theories face, namely, the problem of misconnection, is similar to a suggestion made in Halpern 2015 for evaluating counterfactual dependence in pre-emption situations (the similarity consists in the requirement to hold actual interactions fixed). Halpern, however, is not in the business of developing a process theory but rather, specifying within a structural equation approach the conditions that have to be held fixed to evaluate the conditional counterfactuals relevant for token causation. In Section 4.5 I will briefly discuss the relation of the process theory presented here and the structural equations approach.

account of actual causation? It does not. (1) The account is still an account of processes and interferences into processes. It is by virtue of an interference with a quasi-inertial process, not by virtue of counterfactual dependence, that Suzy's throw is classified as the cause of the shattering – as is illustrated very clearly by the hypothetical pre-emption case in which gravity is switched off. Counterfactuals come into play only when it is necessary to distinguish relevant from irrelevant interferences. (2) The counterfactuals hold by virtue of facts about the quasi-inertial behaviour and the deviations that are described by scientific generalisations (e.g., by exclusive ceteris paribus laws and laws of deviation). The counterfactuals typically *indicate* what the underlying causal structure is.

Fourth, according to my account, the throwing of Billy's stone is not relevant to whether or not the window shatters, but it is certainly relevant to the *manner* of its shattering. The *manner* in which the window shattered would have been different had Billy's stone not been present at all. Thus, we should count Billy's throwing of the stone as an interfering factor – rather than a cause and a pre-empted cause, we would have two contributing causes. The important point is that it depends on the exact effect for which we seek a cause whether my account treats this case as a case of pre-emption or of contributing causes. And that is perfectly fine. If the effect in question is the shattering of the window in contrast to its not shattering, Billy's throw is no cause, while it is a contributing cause to the manner of the shattering. It is thus not true that cases that we intuitively classify as cases of pre-emption have to be classified as cases of contributing causes on my account.

What I suggest here is that the shattering of the window and the particular manner of the shattering are related but numerically different events (or maybe 'aspects' or 'features' of events). One might object that if this kind of proliferation of events is admissible, then there is simply no need to appeal to 'quasi-inertial process' and 'interfering factors' to explain our judgments in the case of late pre-emption; a simple counterfactual theory of causation will suffice. If Suzy had not thrown her stone, the window would not have shattered in the exact manner it did. Rather, an entirely different event would have occurred, one with a slightly different manner of shattering resulting from Billy's stone striking the window. Therefore, the actual effect does counterfactually depend on Suzy's throwing the stone, and now late pre-emption brings no trouble for the necessity of counterfactual dependence for token-causation.

Let me reply by pointing to how what I suggested earlier differs from what Lewis has discussed under the heading of 'extreme fragility'.

According to this approach to the late pre-emption problem, the right way to identify events and, thus, effects is by individuating them such that they could not occur at different times or in a different manner (Lewis 1986, 196–9). If the only way to individuate effects were fragile individuation, then the problem of pre-emption wouldn't occur for the counterfactual account because the actual window shattering would not be the same effect that the throwing of Billy's stone would have caused. However, that is not what I suggest. My suggestion is that we sometimes use fragile individuation and sometimes not. When we don't (e.g., when we consider the shattering of the window) we do have the problem of pre-emption, and my suggestion is to classify the throwing of Billy's stone as an *irrelevant* interference for the non-fragile effect. However, we might also use more fragile individuation and ask why the shattering occurred in this particular manner. In that case the throwing of Billy's stone would be a *relevant* interfering factor. Whether or not an interfering factor is a relevant interfering factor is relative to the effect we are considering. In other words: The problem for the counterfactual approach would only dissolve if the only admissible events were fragile events; however, I admit both fragile and non-fragile events.

Fifth, whether or not the throwing of Billy's stone is a relevant interfering factor, i.e., a cause, depends on the individuation of the effect, as we have seen. By individuating the effect as an effect, we consider it to be a deviant state of a system that would have developed into another state, its quasi-inertial state at that time. If we consider as the effect *the shattering of the window*, we assume that there was a different state of the system (the window being intact) that it would have developed into had nothing interfered. If we consider as the effect *the particular manner of the shattering of the window*, we again assume that there was a different state of the system, its quasi-inertial state (another manner of the window being shattered), that it would have developed into had nothing interfered.

As discussed in Chapter 3, Blanchard and Schaffer criticise default relativity: 'As such default-relativity often seems to us to come close to a free parameter [. . .], which basically gives the theorist leeway to hand-write the result she wants' (Blanchard and Schaffer 2017, 192). A similar remark may be thought to apply to the 'system-relativity' we have introduced. With respect to the example discussed here, one might for instance object: What counts as the cause is highly dependent on how we choose to individuate the effect. Thus, it is presumably true that if we use the rich and detailed description of the effect, then Billy's throwing of the stone does count as (part of) what interferes with the quasi-inertial process (since we know in virtue of physical laws of interaction, conservation of energy, etc., that the window would not have shattered in that specific manner had it not been for the presence of Billy's stone right then and there). So, this system-relativity can generate whatever result the theorist desires willy-nilly, for all we need to do is choose the system *with a description of the effect in however much detail we need* to generate the result we are hoping for – either Suzy's throwing of the stone alone is a cause or Suzy-Billy's throwing together is a cause.

What this objection is missing is that a cause is always the cause of a particular effect. I don't see that there is a problem if the throwing of the stone by Suzy is a cause of the shattering of the window, while the throwing of the stones by both Billy and Suzy comes out as contributing causes to the particular manner of the shattering. So, the relativity that is relevant here amounts to no more than the fact that relative to effect A we have causes of a certain kind, while relative to effect B we have different causes.

Let me stress that the only agent-relativity that plays a role in this account is the one that is generated by the 'explanation seeking why-question' (van Fraassen 1980). Once the effect (the explanandum in the causal explanation, i.e., the effect and its contrast) is fixed, everything else is a matter of quasi-inertial processes and interferences into these processes. Whether or not these facts obtain does not depend on the agent's interest.

Sixth, Paul and Hall raise another worry that is relevant here: Suzy's stone relevantly interfered with the window to bring about its shattering, but so did *Suzy's and Billy's stones taken together*. That would give us the wrong diagnosis because Suzy's and Billy's throwings would both come out as causes of the shattering of the window in the pre-emption case. Thus, the two stones taken together ought to be excluded as a cause of the shattering. Paul and Hall worry that this cannot be done without relying on causal judgments, thus undermining the reductive program (Paul and Hall 2013, 111). As already mentioned, my aim is not to reduce causal facts to non-causal facts. Still, the account presented here needs to be augmented so as to exclude the stones taken together as a cause of the shattering. What needs to be added is some sort of minimality constraint.

As far as I can see, every account of causation needs a minimality constraint, so I could simply argue that I do not need to go into this in any detail. However, a minimality constraint will have to do some extra work on my account of causation when it comes to symmetric overdetermination. So, while there is not enough space for a detailed account of a minimality constraint, I will at least indicate how I would augment the account so as to deal with the problem. First, I will help myself to the notion of a conjunctive event, which is an event that has 'parts' or conjuncts, which are events too. An example would be the conjunctive event consisting of the conjunct events of a throw and a sneeze, or *Suzy's and Billy's throws taken together*, which consists of Billy's throw and Suzy's throw, respectively. Furthermore, I assume that there is (at least in the cases we are interested in) a natural decomposition of such conjunctive events. To minimalise a conjunctive event is to take away at least one of its conjuncts. A maximally minimalised relevant interfering factor for some effect e is a relevant interfering factor that has been shorn of all of those 'parts' without which it is still a relevant interfering factor for some effect e. Suppose Suzy sneezes while she throws a stone at the window. The conjunctive event of throwing and sneezing is a relevant interfering factor but not a maximally minimalised relevant interfering factor. The throw without the sneeze is (presumably) a maximally minimalised interfering factor, while the sneeze without the throw isn't.

So how do we determine whether a relevant interfering factor for a particular effect can be minimalised? We consider whether the effect in question would have been brought about if 'parts' or conjuncts of the relevant interfering factor had been eliminated. So the counterfactual we need in order to test whether we can minimalise a relevant interfering factor is the following: In the antecedent, we consider a situation in which the relevant interfering factor has been shorn of at least one 'part' (all the other interferences are as they are in the actual situation – except those interferences that occur as a consequence of the relevant interfering factor under consideration – and no interfering factors have been added). In the consequent we specify the effect in question. If, despite the elimination of the 'part', the counterfactual comes out as true, the relevant interfering factor can be minimalised. It is important to notice that this is a two-step process. First, we figure out the relevant interfering factors and in a second step these factors are minimalised.

With this notion of a minimality constraint at hand, we can give an augmented version of a definition of a cause (in the disruptive sense):

Let $Z_i (t_2)$ be the state that a system S (at an earlier time t_1) is disposed to be in – provided nothing interferes between t_1 and t_2. Let e be the event of S being in state $Z_e(t_2)$ at t_2, where $Z_e(t_2) \neq Z_i(t2)$.

(DC$_{aug}$) An event c is a cause$_{DC}$ of an event e iff c is the event of some maximally minimalised factor interfering with the quasi-inertial behaviour of system S between t_1 and t_2 such that S at t_2 is in state Z_e, rather than in state Z_i.

or (replacing the phrase 'interfering factor [...] such that" by 'relevant interfering factor'):

(DC$_{aug}$) An event c is a cause$_{DC}$ of an event e iff c is the event of some maximally minimalised relevant factor interfering with the quasi-inertial behaviour of system S between t_1 and t_2 for S at t_2 being in state Z_e, rather than in state Z_i.

Admittedly, all of this is still vague. For instance, the minimality constraint might be too restrictive. Suppose that Suzy's stone is a rather heavy stone. If she had thrown only one half of it, that would also have interfered with the quasi-inertial behaviour of the window and shattered it. Thus, the throwing of the original stone would not count as a cause. I take this to be a serious worry, but I also think that there is a sense in which Billy and Suzy's both throwing, or Suzy's sneezing and throwing, are conjunctive but Suzy's throwing a single stone is not. This is the sort of distinction that seems natural but is difficult to make philosophically precise, but that might fit well with our ordinary language notion of causation.

Be that as it may, what I hope to have indicated is how the minimality constraint might work in cases in which we have clear intuitions.

Finally, symmetric overdetermination: Harry and Sally both throw stones at the window. Unlike in the previous case, the stones reach the window in the very same moment. The windowpane shatters. If only one of the throws had occurred the windowpane would nevertheless have gone to pieces. The general verdict seems to be that both throws should be classified as causes. Traditional process theories have no problem with this case. However, it is my solution to the traditional process theory's problem of misconnection – namely, the additional requirement that causes need to be *relevant* interfering factors – that seems to generate a problem for my account. According to the test for relevance I have introduced earlier, neither Harry's nor Sally's throw is relevant.[5]

However, the minimality constraint, which we have to introduce for independent reasons, as we have just seen, promises to solve this problem. Note that Harry's and Sally's throws taken together is a relevant interference with respect to the shattering of the window. But this conjunct event is not a maximally minimalised relevant interfering factor for the shattering of the window. Indeed, minimalising yields two different maximally minimalised relevant interfering factors: Harry's throw, on the one hand, and Suzy's throw, on the other. In other words, while the non-augmented definition of (disruptive) causation yields the wrong verdict (the

---

[5] Thanks to Sebastian Schmoranzer for pressing this point.

conjunctive event is a cause but its conjuncts aren't), the augmented definition, which incorporates the minimality constraint, classifies Harry's throw as a cause as well as Suzy's throw but not the conjunctive event. This seems to be the right result.

4.2 Disconnections

Dowe (2009) has identified three major problems for process theories. The first issue concerns what has been called the 'problem of misconnection': Some interactions or interferences do not qualify as cases of causation. This problem has been dealt with in the previous section. The second problem is the 'problem of reduction': Process theories that explicate the notion of process and interference in terms of conserved quantities have to assume that all causation is ultimately physical causation (provided only physical quantities are conserved). This problem is not relevant for the account presented here, because quasi-inertial processes and interferences are not defined in terms of conserved quantities that happen to exist only in the physical realm. Finally, there is the problem of negative causation ('the problem of disconnection'), to which I will now turn.

Absences seem to be causally relevant. For instance, the gardener's not watering the flowers caused the shrivelling of the flowers. There are many examples where absences cause, are caused, or are part of a process leading from the cause to the effect. Such 'negative causation' cannot be integrated into process theories that require the persistence of physical characteristics along a world-line that connects cause and effect (see Schaffer 2004 for an extended discussion). This is a problem for some process theories of causation, such as Dowe's and Salmon's, because on their account there is no causation without a physical connection in the sense of transmission of some amount of a conserved quantity (see, e.g., Dowe 2009). On the account involving quasi-inertial processes that I have presented, there is no analogous requirement and, therefore, the problem of disconnection is easier to cope with. Absences or negative events may, on my account, be integrated into quasi-inertial processes. This can be illustrated through the well-known case of double prevention:

> Suzy is piloting a bomber on a mission to blow up an enemy target, and Billy is piloting a fighter as her lone escort. Along comes an enemy fighter plane, piloted by Enemy. Sharp-eyed Billy spots Enemy, zooms in, pulls the trigger, and Enemy's plane goes down in flames. Suzy's mission is undisturbed, and the bombing takes place as planned. (Hall 2004, 241)

Suppose we want to say that Billy's pulling the trigger is a cause of the bombing of the target. Note that there is no continuous physical process leading from Billy's pulling the trigger to the actual bombing, and thus the Dowe–Salmon account has a problem here.

What I want to say about Billy's pulling the trigger as a cause is best presented by building up the quasi-inertial process from simpler cases. Let's start with a very simple situation. Suppose there are neither Enemies nor Billys around and we want to say that Suzy's bombing is the cause for the destruction of the target. In this case, *the target sitting around peacefully* is the quasi-inertial process, which is disturbed by Suzy's interference, i.e., by the bombing. Consider now a second, different case: Enemy is around (but no Billy) and shoots down Suzy. We want to say that Enemy's intervention is the cause of the target's not being bombed (here the effect is a negative event). In this case we consider a different quasi-inertial process, viz., *Suzy bombing the target*. This quasi-inertial process is interfered with because Enemy prevents Suzy from bombing the target.

Finally, add Billy to the picture: Billy shoots down Enemy, and the quasi-inertial process we are considering in this case is even more complex: It is *Enemy preventing Suzy from bombing the target*. So, the system in question consists of the enemy, Suzy and the target. The temporal evolution the system is disposed to leads to a state Z_i (provided nothing interferes), and that is the state in which the target is not bombed because Enemy prevents Suzy from doing so. However, this quasi-inertial process that would lead to Z_i (as specified) is interfered with by Billy's pulling the trigger, such that instead of the state the system was to disposed to manifest (Z_i), another state (Z_e) obtains, namely, the target being bombed. Billy's pulling the trigger is therefore the cause for the quasi-inertial process *Enemy preventing Suzy from bombing the target* not taking place and thus a cause of the bombing of the target.[6] So, even though there is no continuous physical process leading from Billy's pulling the trigger to the actual bombing, on our account we can explicate why Billy's pulling the trigger might be considered as a cause of the bombing of the target. What this case illustrates is that quasi-inertial processes may comprise negative events.

However, negative events may not only be part of quasi-inertial processes; they may also serve as causes or effects. Let me briefly comment on this issue. We have already mentioned a negative event that was an effect,

---

[6] A discussion of the question of whether causation is transitive, which might be relevant for this case, will be taken up in Section 4.4.

viz., the bombing not taking place. When we classify absences (or negative events) as effects we are (implicitly) stating that a certain positive event – the bombing of the target – which would have been part of an undisturbed quasi-inertial process does not take place. Let us finally turn to the case of absences as causes. Consider the case where my neighbour Peter's not watering my flowers while I was on holiday caused the flowers to shrivel. What we are considering as a quasi-inertial process is the flowers living their usual life, with enough oxygen, and watered regularly by Peter so that they flourish. Why is it that the quasi-inertial process is not displayed? Because Peter did not water the flowers. So, we consider Peter's not watering the flowers as the interfering factor. (This case is a little trickier because in this case the interference is not due to an external factor – as is usually the case –rather, the 'interference' is due to some constituent factor of the quasi-inertial process being absent.)

To sum up: Because the account presented here does not require the persistence of physical characteristics, negative causation does not pose a special problem for this kind of process theory.

4.3   More than One Cause

It might seem that the account I have provided only treats a discriminatory sense of causation – i.e., only treats *the* cause of an effect rather than *a* cause of an effect – because it seems to require that a disruptive cause is sufficient for the occurrence of the effect (it requires that a cause is *such that* S at t is in state Z_e, i.e., such that the effect occurs).

However, that appearance is misleading, because causes are required to be sufficient relative to the quasi-inertial behaviour of a system that would have led to the non-occurrence of the effect. There may be more than just one system/quasi-inertial behaviour that has been interfered with. When we attribute more than one cause (*a* cause rather than *the* cause), we consider more than one system (unless we are considering overdetermination cases) and, thus, more than one quasi-inertial process that has been interfered with. When we consider Suzy's bombing to be a cause of the destruction of the target, the quasi-inertial process is either *target sitting around peacefully* or (if we choose a more comprehensive representation) *Billy shooting down Enemy and target sitting around peacefully*. The fact that Suzy's bombing is a cause of the destruction of the target is compatible with there being further causes. Billy's pulling the trigger is another cause. When we consider Billy's pulling the trigger as a cause, the quasi-inertial process interfered with is *Enemy preventing Suzy from bombing the target*.

What is important is that both quasi-inertial processes have actually been interfered with and both interferences are events *such that* (given the relevant quasi-inertial processes) the bombing of the target occurs.

More generally, the account defended here allows for more than one cause because one and the same event (the effect) may be analysed as the result of various actual interferences with different quasi-inertial processes. The destruction of the target was caused by Suzy dropping the bombs in virtue of the fact that this interfered with the quasi-inertial process that would have led to the target remaining undestroyed. The destruction of the target was also caused by Billy pulling the trigger in virtue of the fact that this interfered with another quasi-inertial process that would have led to the target remaining undestroyed: *Enemy preventing Suzy from bombing the target.*

What we have is, thus, an account of causation that allows us to say that an event has *more than one* cause. (However, I am not claiming that all the factors that go into Mill's total cause can be meaningfully captured by this account. Rather, what I called 'closed system causation' seems to be relevant here; see Section 3.8.1.)

4.4 Transitivity

In causal reasoning, we often assume that causation is transitive. We assume schema T: if c is a cause of d, and d of e, then c is also a cause of e. However, there are well-known counterexamples to T, e.g., a case now known as 'boulder' (originally due to Ned Hall):

> A boulder is dislodged, and begins rolling ominously towards Hiker. Before it reaches him, Hiker sees the boulder and ducks. The boulder sails harmlessly over his head with nary a centimeter to spare. Hiker survives his ordeal. (Hitchcock 2001, 276)

So, the rolling boulder causes the hiker to duck, the ducking causes the survival of the hiker, but the rolling boulder does not cause the survival of the hiker. And there are other cases. Causation isn't transitive. Thus, the question arises of whether there is an analysis that can tell us under what circumstances schema T is applicable.

Let us have a look at what goes wrong in the boulder case. Recall our definition of a cause from Section 3.4:

> An event c is a cause of an event e iff c is the event of some factor interfering with the quasi-inertial behaviour of system S between t_1 and t_2 such that S at t_2 is in state Z_e rather than in state Z_i.

For the rolling of the boulder to be a cause of the survival of the hiker, we first have to consider the quasi-inertial process that would have brought about the death of the hiker.

In the scenario we are considering, the death of the hiker would have been brought about by the rolling boulder. Thus, the rolling boulder is a part of the quasi-inertial process that would lead to the death of the hiker. As a consequence, when we are considering the boulder as the cause of the hiker's survival, we view it both as a constitutive part of the quasi-inertial process and as serving as its external disturbing factor. That does not work. The boulder cannot be both a constitutive part of the relevant quasi-inertial process and the process's external interfering factor.

The upshot is that our account of causation presented here allows us to understand why T does not hold in this situation: When we examine whether the rolling of the boulder causes the survival of the hiker, the relevant quasi-inertial process that is disturbed by the boulder and would have led to the death of the hiker cannot be constructed. A necessary condition for c to cause e (provided c causes d and d causes e) is that a system's quasi-inertial processes can be concatenated, such that the concatenated process would have led to the non-occurrence of e but has actually been disturbed by c. If the relevant quasi-inertial process cannot be concatenated, schema T does not hold.

By contrast, such a concatenated system/process can be constructed without running into perplexities in paradigmatic cases in which schema T does hold. Consider the following illustration: Ball 1 bounces into ball 2 (event c). Ball 2 bounces into ball 3 (event d) and ball 3 moves (event e).

Here we want to say not only that c causes d and d causes e, but furthermore that c causes e. But what are the conditions that need to be satisfied such that if c is a cause of d and d is a cause of e, c is also a cause of e?

The essential point is that in the case under consideration a concatenated system S does indeed exist. It is composed of the two balls 2 and 3 taken together. The quasi-inertial behaviour that the compound system

Figure 2 Transitivity I

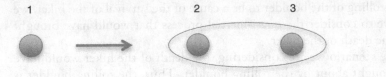

Figure 3 Transitivity II

S is disposed to consists in the two balls not moving at t_e, i.e., at the time when e in fact takes place. This implies in particular that ball 3 is not moving, i.e., that event e does not occur. (Why is there this implication? Because the event of the compound system S displaying its quasi-inertial behaviour (e.g., being in state Z_i at t_e) can be broken down into the two subsystems being in states Z_{i_2} at t_e (ball 2 not moving) and Z_{i_3} at t_e (ball 3 not moving) – here again, I help myself to the notion of a conjunctive event (see Section 4.1).)

The quasi-inertial behaviour of the compound system S is disturbed by ball 1 bouncing into ball 2 (event c) and thus interfering with system S. As a consequence, the quasi-inertial behaviour of the compound system S is not displayed.

Given that c causes d and furthermore that d causes e, we can now explain why, in this case, c also causes e: It is by virtue of the fact that ball 1 causes ball 2 to move (c causes d) that the quasi-inertial behaviour of the compound system S is not displayed. Furthermore, by virtue of the fact that d causes e, the fact that the quasi-inertial behaviour of S is not displayed takes a particular form: It is not only that one of the constituent subsystems, namely ball 2, starts to move, but that the movement of ball 2 causes ball 3 to move as well. So, the event of the compound system S not displaying its quasi-inertial behaviour in this case implies that event e takes place. (Again: Why is there this implication? Because the event of the compound system S not displaying its quasi-inertial behaviour but rather (at t_e) being in state Z_e can be broken down into the two subsystems being in states Z_{e_2} at t_e (ball 2 bouncing into ball 3) and Z_{e_3} at t_e (ball 3 starting to move).) Thus, we can explain in terms of the concatenated system S why if d causes e and c causes d, this implies c causing e.

The essential question for whether or not schema T holds is, thus, whether or not a concatenated system or process that is disposed to a quasi-inertial behaviour such that event e does not occur can indeed be constructed. Remember that in the case of the boulder the process cannot be constructed. The rock cannot simultaneously be a part of a process that is disposed to kill the hiker and the disturbing factor of this process. This analysis can be extended to other cases in which T does not hold, too, as far as I can see. Paul and Hall (2013, 215–16) provide an example that goes back to Hartry Field: Suzy's enemy places a bomb next to her apartment. Billy realises this and pinches out the fuse. However, we don't want to say that the enemy's planting the bomb causes Suzy's survival. Again, as the situation is described, there is no quasi-inertial process that would have led to Suzy's death but is disturbed by planting the bomb next to her apartment. The example has the same structure as the boulder case and thus allows the same diagnosis: The placing of the bomb cannot be both a constitutive part of a process that would lead to the death of Suzy and an external interfering factor for this process.

Another case that has been raised by McDermott and is discussed by Hitchcock is trickier:

> Terrorist, who is right-handed, must push a detonator button at noon to set off a bomb. Shortly before noon, he is bitten by a dog on his right hand. Unable to use his right hand, he pushes the detonator with his left hand at noon. The bomb duly explodes. (Hitchcock 2001, 277)

The dog bite caused the terrorist to push the detonator button with his left hand, which in turn caused the bomb to explode, but we don't want to say that the dog bite caused the bomb to explode. Here is why: As the story is told, it is clear that the person in question is determined to push the button come what may (by using either a hand or a foot or maybe even his head). For the dog bite to cause the bombing, according to our account, there needs to be a quasi-inertial process that would have led to the bombing not taking place. However, as the story is told, the terrorist would have used any means to push the button. Consequently, a quasi-inertial process that leads to the outcome that the button is not pushed would have to be a process with the terrorist not being present (or at least with the terrorist not being able to push the button). But that would not be a quasi-inertial process that the dog bite could interfere with such that the explosion of the bomb is brought about.

Given the account of causation presented in this and the previous chapter, we can explain our causal intuitions both in the cases in which we can use schema T and also in those cases in which it is inappropriate. Furthermore, we can explain what it is that makes schema T inapplicable in certain cases: It is not possible to concatenate a quasi-inertial process that the alleged cause interferes with so as to bring about the event to be explained.

4.5   Causal Modelling and the Role of Quasi-Inertial Processes

Let me finally briefly relate the account of causation in terms of quasi-inertial process and interferences to the recent work in causal modelling.

Causal modelling has become influential as a device for causal inference in the social sciences, psychology, medicine and other disciplines that need to infer from probabilistic correlations to causal structure (Pearl 2000, Spirtes, Glymour and Scheines 2000, Schurz and Gebharter 2016). While causal models as well as causal graphs and the structural equations on which the former rely are, in the first place, a way of *representing* causal structure, in the philosophically oriented literature they have also been invoked to *define* causation (Hitchcock 2001, Woodward 2003, Halpern and Pearl 2005).

I will make no attempt to characterise this approach in any detail (the best introduction is still provided by Woodward 2003). What I am interested in is that various authors working in this framework have attempted to define token causation (actual causation) as part of this overall approach. How does such a definition of actual causation relate to the account outlined in this and the previous chapter?

Within the casual modelling approach, actual or token causation is usually spelled out in terms of conditional counterfactual dependence, where the truth values of the counterfactuals are determined by the structural equations. As an example, consider Woodward's definition of actual causation:

(AC1) The actual value of X=x and the actual value of Y=y. (Woodward 2003, 77)

This first condition simply states that for an event which consists in variable X having the value x to be the cause of the event of variable Y having the value y, both these events need to be actual.

Woodward's second condition runs as follows:

> (AC2*) For each directed path P from X to Y, fix by intervention all direct causes Z_i of Y that do not lie along P at some combination of values with their redundancy range. Then determine whether, for each path from X to Y and for each possible combination of values for the direct causes Z_i of Y that are not on this route and that are in the redundancy range of Z_i, whether there is an intervention on X that will change the value of Y. (AC*2) is satisfied if the answer to this question is 'yes' for at least one route and possible combinations of values within the redundancy range of the Z_i. (84)

The notion of a redundancy range is explained as follows:

> The values $v_1, \ldots v_n$ are in what Hitchcock calls the *redundancy range* of the variables V_i with respect to the path P if, given the actual value of X, there is

no intervention that in setting the values $v_1, \ldots v_n$, will change the actual value of Y.

Crucially, once you invoke the notion of conditional counterfactual dependence by relying on redundancy ranges (or something in the vicinity – the exact details are still debated; for a more recent discussion see, for instance, Halpern 2015), you will get correct results in problematic cases, e.g., in pre-emption cases. The basic idea is that you fix the value of potential back-up causes, which intuitively do not contribute to the effect (such as Billy's stone) in such a way that their non-contribution is held constant. Given that, you check whether the effect depends counterfactually on the other cause (-variable), i.e., Suzy's stone, and it does.

What is essential here is that all the work is done by the redundancy range, which has been added to the causal modelling account so as to yield the correct results regarding our causal intuitions. It has been objected that nothing in the causal modelling framework motivates the addition of this feature.[7] However, there is a rejoinder to this objection that relies on the account of causation in terms of quasi-inertial processes and interferences.

The point to start with is that causal models *represent* underlying mechanisms.[8] A definition of actual causation in terms of causal models such as the one quoted earlier should thus be taken as a prescription *for identifying* what an actual cause is. Given this perspective, it becomes clear that an account of actual causation such as Woodward's is not competing with the approaches discussed so far. Whereas, for instance, process theories give a metaphysical account of what causation is, the casual modelling literature is concerned with identifying causes.

If the causal modelling literature is understood in this way, we can ask what it is that explains the success of this practice of causal reasoning. For instance, why is it that we can successfully identify actual causes given Woodward's recipe? The account of disruptive causation developed in the previous chapters can explain why one should rely on a redundancy range (or something in the vicinity): Satisfying the condition that the values of the other possible causes be in the redundancy range is tantamount to saying that nothing interferes with the process leading to the effect-variable Y having a certain value. In other words, you consider how the system that is characterised in terms of Y would behave provided there are no

---

[7] Paul and Hall when discussing conditional counterfactual accounts of causation formulate the following desideratum: 'What principles determine the selection of the fact F to be held fixed?' (Paul and Hall 2013, 112).

[8] Pearl (2000, xiii–xvi) is very clear about this. Woodward, however, probably wouldn't concur.

interfering factors, that is, its quasi-inertial behaviour. You then consider whether an intervention on X interferes with Y's quasi-inertial behaviour. Thus, in order to get from structural equations to actual causation you have to add – e.g., via the redundancy range – the disruptive concept of causation.

The account of disruptive causation presented here may thus serve as an explanation for why a definition of (i.e., a recipe for identifying) actual causation in the causal modelling framework is successful if it has the form it has. The objection that the definition of actual causation cannot be motivated can thus be countered provided causal modelling is taken to represent causal structure that can be spelled out in terms of quasi-inertial processes and interferences.

CHAPTER 5

Reductive Practices

5.1 Introduction: Reductive Practices and Their Ontological Implications

The general topic of this and the following two chapters is the investigation of some of our reductive practices with respect to the question of what ontological assumptions these practices commit us to. The reductive practices I will focus on are reductive *explanatory* practices as well as the construction of reductive relationships in inter-theory relations. These reductive practices are sometimes taken to be constitutive of or at least to be evidence for fundamentalist claims. The theoretical physicist Max Dresden, for example, observes that 'most physicists would agree that among the sciences physics is surely the most fundamental discipline' (Dresden 1974, 133). He then goes on to explicate what is meant by the claim that one area is more fundamental than another: 'A field A is more fundamental than B, if A explains and describes everything that B does *and more* (original emphasis).' Dresden notes, first, that

> there are a variety of ways in which the 'more' could be interpreted, it could mean 'with greater accuracy' or 'more different phenomena' and, second – and more important in our context – that 'the definition of 'fundamental' just given uses the idea of 'explanation' in an essential fashion. (Dresden 1974, 137)

Dresden then sketches what he means by 'explanation' as follows: For the present purposes an explanation of the behaviour of a system consists of 'the *reduction* of the properties of the system to those of the triple: *constituents, interactions, dynamics*' (Dresden 1974, 137).

The fact that we can explain the behaviour of compound systems in terms of that of the parts is taken to be essential for the fundamentality of physics.

More recently, Carl Hoefer and Chris Smeenk have provided another attempt to spell out physical fundamentalism. They consider what they call the 'received' or 'ideal' view of the physical sciences. Among other claims, this view is committed to fundamentalism:

> Fundamentalism: there is a partial ordering of physical theories with respect to 'fundamentality'. The ontology and laws of more fundamental theories constrain those of less fundamental theories; more specifically: (1) the entities of the less fundamental theory T_i must be in an appropriate sense 'composed out of' the entities of a more fundamental theory T_f, and they behave in accord with the T_f-laws; (2) T_f constrains T_i, in the sense that the novel features of T_i with respect to T_f, either in terms of entities or laws, play no role in its empirical or explanatory success, and the novel features of T_i can be accounted for as approximations or errors from the vantage point of T_f.' (Hoefer and Smeenk 2016, 117)

Hoefer and Smeenk's first condition requires that the behaviour of the less fundamental system can be part-whole explained in terms of that of the parts, i.e., the more fundamental level. In their explication of condition (2) and in particular, of the claim that 'the novel features of T_i can be accounted for' Hoefer and Smeenk write:

> The empirical and explanatory success of T_i must be grounded in the fact that it captures important facts about the structures identified by T_f. Or, in other words, T_i's success should be recoverable, perhaps as a limiting case within a restricted domain, in terms of T_f's ontology and laws. (Hoefer and Smeenk 2016, 117)

So, what Hoefer and Smeenk require for fundamentalism is that the behaviour of compounds can be explained in terms of the behaviour of the parts and that theories that describe that behaviour of the compound can be reduced (recoverable as a limiting case) to the theories describing the behaviour of the parts.

The passages quoted leave it open whether physical fundamentalism as defined by Dresden or by Hoefer and Smeenk comes with ontological commitments. However, some ontological commitments suggest themselves: If physical fundamentalism is true, then physical theories give us *the real story* of what is going on in the world. The claim that physics gives us *the real story* of what is going on in the world can be spelled out in different ways. First, it might be read as an eliminativist claim: The only facts there are, are fundamental physical facts; macroscopic everyday facts are only apparent facts (*Physical Eliminativism*). Second, it might be construed as a claim about ontological priority or metaphysical dependence. According to this reading, the existence of 'higher-order' facts is not denied; there are different layers of facts. On the second reading, the view is spelled out as the claim that all other facts, i.e., the higher-order facts, obtain *in virtue of* or are *metaphysically dependent on* fundamental physical facts. The latter facts are ontologically prior to all other facts (*Foundationalism*).

It may seem natural to suppose that either Physical Eliminativism or Foundationalism is implied by our reductive practices or may be indispensable to make sense of our reductive practices. By contrast, I will argue that this is not the case: neither the assumption of an ontological priority relation nor an eliminativist view is necessary to account for the reductive practices we have. A metaphysics of scientific practice that attempts to be minimal in the sense of not postulating anything beyond what is needed to account for the success of scientific practice does not require the assumption that nature satisfies an ontological priority relation (Chapter 6). Nor will it end up, as I will argue, being committed to Eliminativism (Chapter 7).

Chapter 5 will be devoted to the following two questions: 1) How are the above-mentioned reductive practices to be characterised? 2) Why are we interested in these practices, i.e., what is the *rationale* behind the practice? So, this is not yet the question of how to best explain the success of our reductive practices but, rather, the question of whether the mere fact that reductive practices are something we go for presupposes Foundationalism or Eliminativism. In Chapters 6 and 7, I will turn to the explanation of the success of reductive practices and will investigate in particular, whether our reliance on part-whole relations and on an explanatory backing relation thereby commits us to Foundationalism or Eliminativism.

In the first part of this chapter (Sections 5.2 to 5.4), I will be concerned with disentangling various notions of reduction that have been discussed in the philosophy of science literature and elsewhere. Starting with Section 5.5, I will argue that we are interested in these forms of reduction because we aim at understanding how different accounts of one and the same system can simultaneously lead to successful predictions, manipulations, etc. The rationale for reductive practices, I will argue, is an epistemic commitment that is constitutive of good scientific practice. There is thus no need to assume that the search for theory reduction or for reductive explanations makes sense only on the basis of ontological assumptions along the lines of Foundationalism or Eliminativism.

5.2 Concepts of Reduction[1]

The term 'reduction' is typically used in the sciences and in philosophy both with respect to representations of the world and with respect to features of the world itself. Reduction is often conceived of as a relation between two items of the same kind, e.g., between two properties, states, theories, models or explanations. It is commonly thought of as asymmetric: if an item A reduces to an item B, then B does not reduce to A in the same sense of reduction. My starting point is an analysis of reduction as an asymmetric relation between representational items.

One influential line of thought considers theories to be the relevant representational items. The debate about theory reduction has been shaped by the logical empiricists' views and by Ernest Nagel's in particular. For the logical empiricists the concept of reduction was closely associated with those of *Unity of Science* and of *Progress*:

> The label 'reduction' has been applied to a certain type of progress in science. [. . .] the replacement of an accepted theory [. . .] by a new theory [. . .] which is in some sense superior to it. Reduction is an improvement in this sense. (Kemeny and Oppenheim 1956, 6–7)

Much depends on how this replacement is spelled out in detail. As the authors make clear, it is of particular importance that the new theory can *explain* all the phenomena the old theory could explain, and explanation is explicated in terms of logical derivation.

Some logical empiricists proposed an even stronger claim: The sciences progress via *micro-reduction*: 'We [. . .] think the assumption that unitary science can be attained through cumulative micro-reduction recommends itself as a working hypothesis' (Oppenheim and Putnam 1958, 8). Oppenheim and Putnam characterise micro-reduction as follows:

> The essential feature of a micro-reduction is that the branch B_1 deals with the parts of the objects dealt with by B_2. [. . .] Under the following conditions we shall say that the reduction of B_2 to B_1 is a micro-reduction: B_2 is reduced to B_1; and the objects in the universe of discourse of B_2 are wholes which possess a decomposition into proper parts all of which belong to the universe of discourse of B_1. (Oppenheim and Putnam 1958, 6)

Even though not many authors *explicitly* endorsed micro-reduction, such a view was implicitly very present because the cases that were singled out as paradigmatic could easily be understood to be examples of

---

[1] Sections 5.2 and 5.3 use materials from Hüttemann and Love 2016.

micro-reduction. Nagel's treatment of reduction, for instance, does not mention micro-reduction but his main example of theory reduction nevertheless trades on the part-whole relation: the reduction of thermodynamics to statistical mechanics concerns – among other things – the relation between the molecules that constitute a gas and the gas as a compound.

It is important to highlight the hypothesis that progress is achieved via micro-reduction because it documents that the concept of reduction was closely tied to a number of issues that took a while to disentangle. Thus, the preceding characterisation of (micro-) reduction involves (at least) the following claims:

> (1) Reduction is a type of *progress* towards the *Unity of Science*; (2) Reduction concerns the relation of *theories*; (3) Reduction involves *derivation*; (4) Reduction involves *explanation*; (5) Reduction concerns the *part-whole relation*. It is by no means self-evident that these claims need to be tied together, and as a matter of fact many of the ensuing debates consisted in disentangling the various strands.

As I already mentioned, Nagel did not explicitly propose micro-reduction. His conception was important because it clearly specified a number of formal and informal conditions that need to be satisfied for theory reduction. The reduction of a theory to another theory can, according to Nagel, be modelled on the deductive-nomological account of explanation: ' [. . .] a reduction is effected, when the experimental laws of the secondary science [. . .] are shown to be logical consequences of the theoretical assumptions [. . .] of the primary science' (Nagel 1961, 352).

If the experimental laws of the old theory can be derived from the new theory, and explanation is conceived of in terms of derivation, then the new theory accounts for all the explanatory successes of the old theory. Furthermore, it might also be superior because it might explain more than the old theory. The old theory, on this account, turns out to be a special case of the new theory.

For Nagel's account to work, two formal conditions have to be met. The condition of connectability requires that the terms of the two theories in question can be connected. In the case of so-called heterogeneous reduction, the theory to be reduced (e.g., thermodynamics) contains expressions (e.g., temperature) that do not occur in the reducing theory (e.g., statistical mechanics). These expressions need to be linked to expressions in the reducing theory (via so-called bridge laws) to ensure that the laws of the reduced theory can be derived from those of the reducing theory (condition of derivability).

If Nagel reduction succeeds, the old theory is completely integrated/ embedded into the new theory. Examples (besides the case of thermo-dynamics and statistical mechanics) that Nagel thought might fit this characterisation are the reduction of Galileo's 'theory' of free-falling bodies as well as Kepler's laws to Newtonian mechanics. These seem to be nice examples of progress by integration of old theories into a new theory.

Nagel reduction is reductive in an intuitive sense of 'getting along with less than we started with' because the principles or axioms of the old theory turn out to be derivable and thus are not needed as axioms. We get along with fewer principles.

While the logical empiricists eschewed realistic interpretations of theories, once one gives up this attitude, successful Nagel reduction naturally leads to an ontological reading of bridge laws: Given a realistic reading of theories, the reduction of, say, thermodynamics to statistical mechanics seems to imply that a bridge law that links, e.g., temperature and mean kinetic energy should be read as asserting that temperature is identical to and nothing over and above mean kinetic energy.

Most of the ensuing debates about reduction proceeded by criticising various aspects of Nagel's account. This led, in particular, to the disen-tangling of various strands that had been lumped together in the logical empiricists' conception of reduction (Hüttemann and Love 2016). For the purposes of this chapter, I will focus on two strands: on how the debates on theory reduction (Section 5.3) and on part-whole explanations (Sections 5.4 and 5.5) evolved.

5.3 Limit-Case Reduction

One strand in reduction debates concerns the question of how different theories with overlapping domains of application are related to each other. Nagel's idea to explicate this relation in terms of logical derivation was criticised early on by Feyerabend (1962). An essential point of Feyerabend's criticism can be illustrated by comparing the predictions of classical Newtonian mechanics (NM) and the special theory of relativity (STR) with respect to the momentum of a massive particle. The problem is that when one approaches the velocity of light (speed $v/c = 1$) the predictions with respect to the momentum diverge significantly.

This is a severe problem for Nagel's account of reduction, i.e., for his account of how special relativity theory as the successor theory is related to classical Newtonian mechanics. According to Nagel reduction, the predictions of classical Newtonian mechanics (or its experimental laws) should be derivable from special relativity. However, because the predictions of the two theories are incompatible, such a derivation is impossible.

By way of rejoinder, one might concede that the old theory typically cannot be deduced from the succeeding theory. Schaffner (1967, 1969, 1976, 1993) as well as Dizadji-Bahmani, Frigg and Hartmann (2010) have argued that a *suitably corrected version* of the old theory should be the target of a deduction from the succeeding theory. The corrected version of the old theory needs to be strongly analogous to the original higher-level theory, where 'strong analogy' is spelled out, for example, as 'good approximation' (Dizadji-Bahmani, Frigg and Hartmann 2010). However, this cannot serve as a general account because, to return to our example, close to the velocity of light, classical mechanics is in no sense strongly analogous or a good approximation to anything that might be derived from special relativity. The problem that was raised by Feyerabend is not merely an accidental feature that turns up in a limited number of cases. Rather, it is a characteristic feature of theory succession, because new theories often make *better* predictions (in some domains of application) and thus are in conflict with the old theory. Nagel's account of reduction as well as attempts to refine it cannot serve as a general model of how succeeding theories are related.

Even though it is false that predictions from classical Newtonian Mechanics (NM) and the Special Theory of Relativity (STR) are approximately the same, it is nevertheless true that *in the limit of small velocities*, the predictions of STR approximate those of NM. While Nagel reduction does not describe the case of NM and STR adequately, there is a different approach that is more promising. As Thomas Nickles pointed out, physicists use a different notion of reduction according to which STR reduces to NM under certain conditions:

> Philosophers and scientists often speak of reducing less fundamental theories to more fundamental ones. (e.g. physical optics to electromagnetic theory, phenomenological thermodynamics to statistical mechanics). But in the case of [NM] and STR it is more natural to say that the more general STR reduces to the less general [NM] in the limit of low velocities. Epitomizing this intertheoretic reduction is the reduction of the Einsteinian formula for momentum,
>
> $$p = m_0 \, v \, / \, \sqrt{(1 - (v/c)^2)}$$
>
> (where m_0 is the rest mass), to the classical formula $p = m_0 \, v$ in the limit as $v \rightarrow 0$. [. . .] This use of 'reduces to' is not only intuitively natural; it is the

way physicists and mathematicians, in contrast to philosophers, usually talk.
(Nickles 1973, 182)

Nickles refers to this latter concept as *limit-case reduction* in order to
distinguish it from Nagel's concept of theory reduction. Limit-case reduction often permits one to ignore the complexities of STR and work with
the simpler theory of NM, given certain limiting conditions.

Limit-case reduction is very different from Nagel reduction, as Nickles
himself already pointed out. The first difference concerns the direction of
reduction. Both concepts describe reduction as an asymmetric relation and
thus as allowing reduction in one direction only. According to Nagel, the
less general theory is reduced to the more general one (the term 'reduction'
is used in virtue of needing *a smaller number of fundamental laws* or basic
assumptions). In contrast, in limit-case reduction, the more general theory
is reduced to the less general one (the term 'reduction' is used in virtue of
needing *less complexity* in the laws). This shows that the asymmetry that
characterises most reduction concepts need not tell us something deep
about the relation of theories or about the relation of the facts described by
these theories. Whether a theory T_1 reduces to a theory T_2 or vice versa
depends on whether we focus on one sense of reduction or the other. This
observation will be relevant in the next chapter because it indicates that the
mere fact that a reduction relation is asymmetric does not have to be read as
being evidence for an underlying asymmetric ontological relation.

A second important difference is that limit-case reduction is a much
weaker concept. Successful Nagel reduction, shows that the old theory can
be embedded entirely in the new theory, whereas limit-case reduction
focuses on two theories that make different predictions about phenomena,
which nevertheless converge under special circumstances. Thus, limit-case
reduction is typically piecemeal – the theories in question may make
contact in some domains of application only. It might be possible for
one pair of equations from STR and NM to be related by a limit-case
reduction, while another pair of equations fails.

Third, even though both accounts involve derivation, they differ in what
is derived. On Nagel's account, the laws of the old or higher-level theory
have to be logically deducible from the new theory. For limit-case reduction, the classical equation (or law) is derived from the STR equation in the
following sense: 'derivation' refers to the process of obtaining a certain
result by taking a limit-process. So, strictly speaking, it is not the classical
equation that is logically derived from STR, but, rather, solutions of the
new equations that are shown to coincide with solutions of the old

equations in the limit, and solutions of the new equations are shown to differ from those of the old theory only minimally in the neighbourhood of the limit (e.g., Ehlers 1997).

There are a number of detailed studies of the exact limiting relations between different theories (Sklar 1993; Uffink 2007 for the relation of thermodynamics and statistical mechanics; Earman 1989; Ehlers 1986; Friedman 1986; Scheibe 1999 for the relation between Newtonian mechanics and the general theory of relativity; and Scheibe 1999; Landsman 2007 for the relation of classical mechanics and quantum mechanics) and a number of problems have emerged, which I will not discuss here (see Hüttemann and Love 2016 for a short overview).

For the purposes of this chapter, the most important issue is the rationale of limit-case reduction. Limit-case reduction is valued because we want to understand the empirical successes of the old theory with respect to prediction and explanation from the perspective of the new theory. Nickles observes:

> Rather than to effect ontological and conceptual consolidation the main functions of [limit-case reduction] are justificatory and heuristic. The development of new theoretical ideas is heuristically guided by the requirement that these ideas yield certain established results as a special case (e.g., in the limit), and they are often quickly justified to a degree by showing that they bear a certain relation to a predecessor theory. It was an important confirmation of STR to show that it yielded [NM] in the correct limit. (Nickles 1973, 185)

Limit-case reduction aims to explain the past success as well as the continued application of a superseded theory from the perspective of a successor theory.[2] Kemeny and Oppenheim had already argued much earlier that it is essential for reduction (or progress) 'that the new theory should fulfill the role of the old one, i.e., that it can explain (or predict) all those facts that the old theory could handle' (Kemeny and Oppenheim 1956, 7). It is, to use a phrase by Rohrlich (1988, 305; Rohrlich and Hardin 1983, 604), a *coherence requirement* that the successes of the old theory should be recoverable from the perspective of the new theory.

To sum up: It is part of good scientific research to search for an account of why a superseded theory was successful in the past and why we should continue to apply it. Limit-case reduction provides answers to these questions.

---

[2] For a more detailed discussion of the functions of reduction in science, see Crowther 2020.

Note that what is explained in limit-case reduction is the *successes* of the old theory. This is not the same as an explanation in a deductive-nomological sense of the *experimental laws* of the old theories as in Nagel reduction (though the latter arguably implies the former). As already indicated, explaining the successes of the old theory does not imply that the old equations or laws are logically derived from the new, but only that the solutions of the new equations are shown to converge to the solutions of the old equations in some limit.

For the purposes of this chapter, this brief sketch will suffice. I will take up some of the issues discussed here from Section 5.6 onwards.

5.4   Part-Whole Explanations

Even though – as we have seen – within logical empiricism, the question of how the behaviour of compound systems can be explained in terms of behaviour of the parts has surfaced in the context of micro-reduction, the issue was subsumed under the heading of theory reduction and not much attention was paid to it. This changed during the second half of the twentieth century, in particular, when it became clear that the Nagel model of reduction was not applicable to reductive practices in disciplines like biology because, e.g., there were no explicit theories to start with (Hüttemann and Love 2011). As a consequence, some authors have conceptualised reductionism, which they assumed to have a place in biology, in terms of the relationship between parts and wholes (e.g., Bechtel and Richardson 1993; Sarkar 1998; Wimsatt 1976). The reductive practice that came under consideration focused on explanations and on the question of what turns an explanation into a reductive explanation. In what follows, I will assume that it is possible to separate the issue of how to characterise an explanation and the issue of whether an explanation is reductive.

A part-whole explanation explains the behaviour of systems in terms of compositional features rather than in terms of the causal history. When we confine ourselves to physics, the *behaviour* of systems will be either the *state* of a system or its *dynamics* (temporal evolution). An example for the explanation of the former is the explanation of the kinetic energy of a gas in terms of the kinetic energies of its constituent parts.

In this section, I will discuss an example for the latter, i.e., for a part-whole explanation of the *dynamics* of a system in detail. I will come back to the part-whole explanation of the state of a compound system in Section 5.6.3.

The dynamics (temporal evolution) of a compound system can often be explained in terms of the dynamics of the parts and their interactions. Both the explanandum and the explanans have a temporal dimension, because the dynamics determines how the state of a system evolves. That, however, does not imply that the explanation is causal. As we will see, no appeal to the causal history of the systems is made. The explanation is purely constitutional or part-whole. As an illustration, I will describe the classical treatment of an ideal crystal and its properties. According to standard textbook accounts, the electrons and ions that constitute the crystal can be considered separately (adiabatic approximation). The regular structure of the crystal is generated by the ions. According to the harmonic approximation, the ions perform oscillations around the sites of the lattice, which are described as the mean equilibrium positions of the ions. These oscillations are considered small in comparison with the inter-ionic spacing, which means that only nearest-neighbour interactions are relevant. Furthermore, it is supposed that the potential between nearest neighbours is harmonic (Ashcroft and Mermin 1976, 422–7). On the basis of these assumptions, we can specify the classical Hamiltonian function of the ideal crystal. The Hamiltonian function is constructed, on the one hand, in terms of the dynamics of the constituents, which are understood as isolated (kinetic energy terms), and on the other hand, in terms of their interactions (potential energy). These contributions are added together:

$$H = \Sigma_i \, E^i_{kin} + (1/2) \, \Sigma_{ij} U_{ij} q_i q_j \tag{5.1}$$

where $E^i_{kin} = p^2_i / 2m$ is the kinetic energy of the parts (i.e., the ions), and $U_{ij} = \partial^2 / \partial q_i \partial q_j \, U(q_1 \, .. q_{3 \, N})$ describes the interactions between the parts (where the q_i are generalised coordinates).

Equation (5.1) contains the core of the part-whole explanation. The Hamiltonian function H for the compound characterises the temporal evolution of the compound system, i.e., of the ideal crystal. This Hamiltonian function is determined by three features, which are constitutive for the part-whole explanation: characterisations of (i) how each part (i.e., each ion) would behave on its own (the kinetic energy term) – i.e., the ions' quasi-inertial behaviour, to use the terms introduced in Chapter 3, (ii) how the ions interact (potential energy term) and (iii) how these contributions are added up.

Given the part-whole explanation of the dynamics of the ideal crystal, its macroscopic properties can be explained or calculated. On the basis of the Hamiltonian function, we can determine the thermal density of the crystal, which is given by

$$u = 1/V \, (\int d\Gamma \, exp \, \{- \beta H\} H)/(\int d\Gamma \, exp \, \{- \beta H\}) \qquad (5.2)$$

where $d\Gamma$ stands for the volume element in crystal phase space and $\beta = 1/k_B T$, where k_B is the Boltzmann constant and T the temperature.

The thermal density of the crystal permits us to calculate the observable behaviour of the compound system, including thermodynamic properties such as the specific heat c_v:

$$c_v = (\partial/\partial T) \, u$$

Let me make a few remarks about part-whole explanations.

First, part-whole explanations can be classified as reductive in (at least) two respects:

(a) The explanation refers exclusively to the parts, their properties, their interactions and laws of composition. By definition, the context of the system in question is considered to be irrelevant – it can be ignored. (In Hüttemann and Love 2011, we called this feature 'intrinsicality'.)

(b) Very often the properties of the parts are of a particular restricted kind, which are considered fundamental and exclude certain other properties of the kind to be explained. In the example, the parts are characterised exclusively in terms of *non*-thermodynamic properties (see Eq. 5.1). This is compatible with the fact that the compound is characterised in thermodynamic terms (specific heat, temperature). The latter terms are connected with the description of the components in part via Eq. 5.2. (In Hüttemann and Love 2011 the feature that the parts are characterised in terms of a restricted set of properties was called 'fundamentality').

While (a) is essential, to part-whole explanations, (b) isn't. Thus, a part-whole explanation may be reductive in satisfying constraint (a) but fail to be reductive in the sense of satisfying (b). A simple example would be the explanation of why a compound has a certain mass value in terms of the mass values of the constituents. (A more sophisticated example will be discussed in Section 5.6.3.)

Second, part-whole explanations may, but need not, invoke theories. The behaviour of the parts and the compounds may be described in terms of local generalisations in the absence of full-fledged theories. This is often the case for part-whole explanations outside physics.

Third, part-whole explanations play a prominent role in debates about emergence. C. D. Broad – one of the most influential figures in debates that took place in the first half of the twentieth century – characterised emergence in terms of the failure of part-whole explanations. (I will come back to this issue in the next section.)

5.5 Failures of Part-Whole Reduction and the Requirement of Micro-Macro Coherence

In this section, I want to argue that in scientific practice, failures of part-whole explanations are considered to be *anomalies* in the sense of Kuhn. According to Kuhn, 'novelties of fact' and 'novelties of theory' are anomalies if 'they violate the paradigm-induced expectations that govern normal science' (Kuhn 1996, 52–3). So, what is characteristic for anomalies is that that they appear to contradict established theories (see also Hoyningen-Huene 1993, 224). Anomalies generate friction and are unacceptable in the long run. Just as there is the constraint to avoid theories that generate false predictions, I argue that failures of part-whole explanations are considered anomalies for epistemic reasons: they violate an epistemic constraint that I will call 'micro-macro coherence'.

I will start by presenting a few examples from the history of science that are meant to illustrate these claims.

Case 1: *The Cartesian program of explaining the behaviour of systems*: Descartes characterised his own account of physical phenomena towards the end of his *Principles*: 'I have considered the shapes, motions and sizes of bodies and examined the necessary results of their mutual interaction in accordance with the laws of mechanics [. . .]' (Descartes 1985, 286). Descartes attempted to explain the behaviour of all compound systems in terms of the size, figure and motion of their parts (plus laws of collision). Together with some additional assumptions, e.g., about the nature and relation of space and matter, this committed him to a vortex theory of planetary motion, according to which the planets are carried in a vortex around the sun (for a detailed discussion see Gaukroger 2002, 142–60 or Schuster 2013, 73–8).

Newton in his *Principia* came to reject the vortex theory: 'The hypothesis of vortices is beset with many difficulties.' One of the problems Newton focuses on is comets because their motion appears to be unaffected by the vortices through which they move: 'The motions of comets are extremely regular, observe the same laws as the motions of planets, and cannot be explained by vortices. Comets go with very eccentric motions into all the parts of the heavens, which cannot happen unless vortices are

eliminated' (Newton 1999, 939). Even though there ensued a long tradition
to save the vortex theory because it was considered simpler and did not
postulate gravitation, which some took to be an 'occult force' (Gaukroger
2002, 159–60), ultimately Newton's rejection of the vortex theory and his
theory of gravitation were accepted.

In other words: Descartes' attempt to explain the behaviour of
a particular compound system, namely, the solar system (plus comets) in
terms of his physics, i.e., in terms of size, figure and motion of their parts
(plus laws of collision) – that is, his particular version of a reductive
explanation – did not succeed. The failure of this reductive project could
have been taken to indicate that there is no explanation of the behaviour of
the compound system (the planetary plus comet system) at all. Thus, one
theoretical option would have been to say: 'Gravitational phenomena
(such as the observed paths of the planets and comets) cannot be explained
in terms of the parts' properties and the laws that apply to them. This is
a case of *emergence*. And that is what we have to accept.'

Why was emergence not a relevant option? Let me briefly elucidate the
notion of emergence that is relevant in this context. C. D. Broad characterised emergence as the failure of part-whole explanations. According to
Broad (1925), the behaviour of a system is emergent if two conditions are
met: (a) the behaviour of the parts *determines* that of the compound, but
(b) the behaviour of the compound *cannot be part-whole explained*, where
part-whole explanation is spelled out as follows:

> A compound system's behaviour is part-whole explainable if it is – at least in
> principle – possible to deduce (to explain) the behaviour of the compound
> on the basis of
>
>   (i)   general laws concerning the behaviour of the components considered
>   in isolation
>   (ii) general laws of composition
>   (iii) general laws of interaction

What is essential for the context of this section is the requirement that the
laws be *general*: this is meant to exclude trivial part-whole explanations.
Suppose there is no explanation of why water is transparent at room
temperature on the basis of what we know about the parts, their interactions, etc. Without the generality requirement, it is easy to make up an
explanation, or at least a deduction, by adding the law that whenever
H_2O molecules come together in suitable numbers, the compound will
be transparent. This is a non-general (and ad hoc) law because it pertains
only to the kinds of systems that are under investigation. The generality

requirement in conditions (i) to (iii) excludes such ad hoc manoeuvres. It is thus a methodological consideration that motivates the generality requirement.

To return to our example: Accepting that gravitational phenomena are emergent in the sense outlined amounts to giving up on reductive explanation. But as a matter of fact, emergence was not seriously entertained as a way out. Rather, Newton and other physicists redefined the basis for a (possible) part-whole explanation. The constituents were now conceived of as possessing an additional property: mass, along with an additional law that describes the relevance of this property, i.e., the law of gravity.

Clearly the original reductionist hypothesis had to be given up. But in hindsight this is not a case of emergence, because, given the redefined basis, the behaviour of the compound can be explained in terms of the properties of the parts, their interactions, etc. (see Hoyningen-Huene 1994 for discussing this example in this context).

So, in the case just considered, one putative reductive explanation is replaced by a different (presumably more successful) reductive explanation. The constraints on the explanans have changed and, thus, so did the conditions that have to be satisfied for an explanation to count as reductive.

To claim that the compound's (solar system's) behaviour is emergent would have been considered to be a mere ad hoc move, which merely redescribes the fact that we have an anomaly in Kuhn's sense. At best, we would have gained an account of the relation of parts and wholes in terms of non-general or ad hoc laws, and that would have been considered to be methodologically problematic.

Case 2: *The specific heat of polyatomic gases*: In the late nineteenth century, there was no consistent mechanical model that would account for the observed specific heats of polyatomic gases. According to Maxwell, 'Here we are brought face-to-face with the greatest difficulty which the molecular theory has yet encountered' (quoted in Darrigol and Renn 2013, 772–3). Again, reduction seems to fail. One option would have been to argue: 'Specific heats of polyatomic gases cannot be explained in terms of the parts' properties plus the laws that apply to them. This is a case of *emergence*. And that's fine.' Again, this option was not taken seriously. The specific heats of the polyatomic gases were considered to be an *anomaly*. Eventually, with the advent of quantum mechanics, the specific heats of polyatomic gases turned out to be part-whole explainable (see Darrigol and Renn 2013).

So, again, we have a case in which the basis for a part-whole explanation was redefined. The constituents were considered to obey quantum statistical mechanics rather than classical statistical mechanics. The constraints on the reductive explanans have changed. One candidate for a reductive explanation is replaced by a different reductive explanation rather than by a non-reductive explanation. To repeat: Not to have a general account of how the behaviour of the parts is related to the behaviour of the compound is considered to be an *anomaly* in the sense of Kuhn.

To claim that the gas's behaviour (i.e., its specific heat) is emergent, is merely a re-description of the situation. At best, it gives us an account of the relation of parts and wholes in terms of non-general or ad hoc laws such as 'Every time the microstate of the gas is such and such the associated specific heat is this and that.' Such an approach would have been methodologically dubious.

Case 3:   *Entropy*: The development of the relation between thermodynamics and statistical mechanics can be taken to illustrate this point further. Boltzmann worked out a mechanical interpretation of the second law of thermodynamics based on probability considerations. Loschmidt and Zermelo confronted Boltzmann with two objections, the reversibility objection (*Umkehreinwand*) and the recurrence objection (*Wiederkehreinwand*) (for a detailed account see Darrigol and Renn 2013). These objections could only qualify *as objections* because of the underlying assumption that the representation of the macro-behaviour of a system (in terms of thermodynamics) and the representation of the behaviour of the parts (in terms of statistical mechanics) have to fit together in the following sense: We want to understand the behaviour of the compound (in this case, its increase of entropy over time) in terms of the general laws that govern the behaviour of the parts, i.e., in this case, the behaviour of the molecules. As long as such an explanation has not been found, we are confronted with an anomaly.

Case 4:   *The protein-folding problem*. Let me finally mention a biological example. In biology, the systems we are interested in are usually embedded in a context, while – at least traditionally – in physics, there was a focus on isolated systems.

Nearly all proteins adopt a three-dimensional structure (tertiary structure) in order to be functional. What is called the 'protein-folding problem' concerns explaining how this active conformation is achieved for polypeptides subsequent to translation from RNA in the cellular context. 'A functional protein is not *just* a polypeptide chain, but one or more polypeptides precisely twisted, folded, and coiled into a molecule of unique shape' (Campbell and Reece 2002, 74). The linear sequence hypothesis holds that the three-dimensional folding of a protein results from the properties of the amino acid residues in the polypeptide and their interactions alone. The folded protein should be explained exclusively by the chemical interactions of its component residues as ordered in a linear polypeptide. The linear sequence hypothesis thus claims that there is a part-whole explanation for protein folding.

As it turns out, an explanation of the protein folding cannot rely on intrinsic features of the amino acid chain alone, i.e., on the behaviour of the parts and their interactions. 'The manner in which a newly synthesized chain of amino acids transforms itself into a perfectly folded protein depends both on the intrinsic properties of the amino acid sequence and on multiple contributing influences from the crowded cellular milieu' (Dobson 2003, 884). The intrinsic properties of the linear polypeptide arising from its amino acid residue parts are not sufficient to explain the protein-folding manifestation in the cell. The process of folding requires not only appropriate environmental conditions but also the contribution of extrinsic proteins, so-called chaperones.

The original reductive hypothesis according to which there is a part-whole explanation for the folding of the protein has failed. However, the folding was not classified as a case of emergent behaviour. It would have been methodologically problematic to account for the relation between parts and wholes in terms of specially tailored ad hoc laws. Rather, the boundaries of the system have been redrawn: Additional factors ('chaperones') have to be taken into account. The charge of emergence can be avoided by blaming the context or by redrawing the boundary (for a more detailed account, see Hüttemann and Love 2011).

What is suggested by these cases is that with respect to part-whole explanations, a failure of reductive explanation (in terms of general laws) will not easily be accepted. It would be considered an *anomaly* in Kuhn's sense. If a reductive explanation fails, there is pressure to come up with a different explanatory account in terms of general laws of how parts and wholes are connected.

Once we have identified certain entities as parts of a compound system, the question arises of whether and how our accounts of the parts and our accounts of the compound are related. It is a constitutive feature of scientific research to achieve what I will call for brevity's sake 'micro-macro coherence':

The representations of the compound system and the representation of the
parts and their interactions should (i) yield the same predictions with respect
to the behaviour of the system in question and (ii) account for why they
yield the same predictions in terms of general laws.

Micro-macro coherence is furthermore important because we need to
understand how an intervention on a part affects the compound and,
conversely, how an intervention on the compound affects the parts not
only on a case-by-case basis, but in general terms. For such an understanding, we need a detailed general account of how what we know about the
parts is connected with what we know about the compound, i.e., we need
a part-whole explanation. Without such a reductive explanation, there
would be no explanation of how the behaviour of the parts is related to
the behaviour of the compound: We therefore require that the laws
concerning the behaviour of the parts and their interactions explain or
account for the laws for the behaviour of the compounds in general terms
without recourse to ad hoc laws.

The fact that in scientific practice, we are looking for reductive explanations and that failures are considered as anomalies indicates that it is an
integral part of what we take to be good scientific practice to search for
general accounts of how the behaviour of parts and wholes is related.
Micro-macro coherence and, thus, the search for general part-whole
explanations are constitutive of good scientific research. In this sense, the
search for reductive explanations functions as a regulative principle in
scientific practice.

The foregoing consideration should not be read as an a priori argument
for the existence of part-whole explanations. Rather, it is an inductive
hypothesis based on past science: The argument can be construed as a kind
of pessimistic meta-induction from the perspective of the emergentist.
Whenever the emergentist might have thought that they had made the
case for emergentism, it turned out that there was an unconceived hypothesis about how the behaviour of the parts, their interaction and the
behaviour of the compound are related (see Stanford 2006 for the notion
of an unconceived hypothesis). What seemed like a good candidate for
emergent behaviour turned out to be reductively explainable in terms of
a new hypothesis about the constituent parts and their interactions.

Candidates for part-whole explanations that fail to explain the behaviour of a compound can (often) be replaced by another part-whole explanation. Maybe this is less surprising than it initially appears. The schema for
part-whole explanation can be read as a reductive explanatory strategy,

which is to some degree flexible. With the exception of the requirement that the laws that go into a part-whole explanation be general, there is no initial constraint on the kinds of properties of the parts or the kinds of interactions. As a consequence, the reductive strategy is flexible in various respects:

1) We can revise our assumptions about the parts' properties. Mass and charge have been added to the repertoire of fundamental physical properties (see, e.g., case 1).

2) We can revise our assumptions about the interactions between the parts. Gravitation and electromagnetism were accepted as real physical forces at some point in the history of science. (The introduction of forces goes hand in hand with the introduction of 'charges', but the latter do not determine the exact form of the force laws.) (See, e.g., case 1.)

3) We can revise our assumptions about the dynamical laws that pertain to these properties. An example is the replacement of classical mechanics by quantum mechanics, which allowed the reductive explanation of the specific heats of the polyatomic gases (see, e.g., case 2).

4) We can re-describe the macroscopic behaviour to be explained within the limits defined by experimental accuracy. (In the case of the second law of thermodynamics, the original phenomenological law was re-described as a statistical law, so as to allow an explanation is terms of statistical mechanics.) (See, e.g., case 3.)

5) Furthermore, in order to avoid the conclusion that the behaviour of a compound system is emergent, one can – as is common in biology – retract the initial assumption that the system under consideration is indeed isolated and redraw the boundaries of the system (see, e.g., case 4).

So, if a particular reductive hypothesis turns out not to work there are a number of options to generate revised hypotheses. Clearly, this in-principle flexibility is constrained by the requirement that there be independent empirical evidence for the revised assumptions that go into such an explanation (at least in cases 1–3 and 5).

To put into perspective what I have argued for in this section, I will point to a number of claims, which are *not* implied.

First, it is not implied that we have actually found part-whole explanations for the behaviour of all kinds of systems, nor is it claimed that this is *in principle* possible. All I am claiming is that the absence of a general account of how what we know about the parts fits together with what we know about the compound would be considered an anomaly that needs to be dealt with.

Second, the claim that the search for part-whole explanations is constitutive of what counts as good science does not imply any sort of methodological reductionism according to which the best and only way to do science is to go for part-whole explanations. All I am claiming is that *if* we have an account of the behaviour of the compound and an account of the behaviour of the parts, *then* a part-whole explanation is *one* thing (among others) that science cares for. There are other kinds of explanations. There are causal explanations. There are explanations that invoke symmetries and ones that use renormalisation group techniques.

Third, I am not committed to the view that there is only one way of decomposing a system. There may be different decompositional strategies, and one and the same system may even be decomposed differently for different purposes (for discussion see Healey 2013). To repeat: All I am claiming is that *once we have* a description of the parts as well as of the compound, these should fit together.

Finally, trying to achieve general accounts of how the behaviour of parts and that of the whole are related is compatible with the claim that it is impossible to develop or generate higher-level descriptions or explanations of higher-level behaviour exclusively by micro-level information. It may very well be the case, to use a phrase by Robin Hendry (personal communication), that macro-level descriptions need to be 'projected downwards', i.e., we might need to rely on macro-level descriptions to delineate the relevant micro-level systems.

All that is claimed is that *once we have* a macroscopic description of a compound system plus a microscopic description (i.e., a description of the constituents), we want to understand how these descriptions are related, which usually leads to part-whole explanations.

5.6 Objections, Counterexamples and Replies

I have argued that the quest for micro-macro coherence is part of what we consider good science and that a failure of part-whole explanations would be considered an anomaly. The search for part-whole explanations is a regulative principle of scientific practice. However, it might be objected that we have settled for the failure of part-whole explanations – in at least some cases we seem to accept emergent behaviour.

By way of answering this objection, let me first point out that taking the search for part-whole explanations to be a regulative principle of scientific practice does not commit me to the claim that we are always successful in constructing such an explanation. However, second, I will argue that – contrary to first appearances – even in those cases that might be considered as counterexamples to various forms of reduction, it is not at all obvious that micro-macro coherence has been given up. I will consider three such cases: phase transitions, decoherence and quantum entanglement.

5.6.1 Phase Transitions

A case that might be thought of as an illustration that we sometimes accept the failure of part-whole explanations concerns phase transitions and so-called critical phenomena, which occur in the vicinity of certain types of phase transitions (see Reutlinger and Saatsi 2018 for a recent discussion).

Thermodynamically, phase transitions and critical phenomena are associated with non-analyticities in a system's thermodynamic functions (i.e., discontinuous changes in a derivative of the thermodynamic function). Given certain uncontroversial assumptions, such non-analyticities cannot occur in finite systems (cf. Menon and Callender 2013, 194) in the canonical or grand-canonical ensembles of statistical mechanics. For phase transitions to occur and for systems to exhibit critical phenomena, it thus appears that an 'ineliminable appeal to the thermodynamic limit and to the singularities that emerge in that limit' (Batterman 2011, 1038) is required.

Why might phase transitions and associated phenomena be considered as counterexamples to the claim that a quest for micro-macro coherence is written into scientific practice? It may be argued that the mere fact that descriptions in terms of statistical mechanics can approximate phase transitions as defined in thermodynamics in the thermodynamic limit (i.e., for infinite systems) does not suffice for the reductive projects we are interested in. The reason is this: We want to understand the behaviour of the compound in terms of its parts. Real systems have only finitely many parts. A system with finitely many components cannot give rise to phase transitions as characterised in thermodynamics. It remains unclear how the idealised description in terms of statistical mechanics for the infinite system pertains to the finite system. It thus seems that we do not have an account of how the behaviour of the parts (characterised in terms of statistical mechanics) and that of the compound (characterised in terms of thermodynamics) fit together.

This is again a case in which the behaviour of the compound and that of the parts are described by different theories. So, the issue of how the two theories are related arises as well: We can only understand the success of thermodynamics in describing (some aspects of) phase transitions and the reason why it is legitimate to continue using it if the idealised statistical mechanical description for the infinite system can be shown to be relevant for the finite system too.

It is uncontroversial that an appeal to infinite idealisation *need not* be a problem if one wants to account for the successes of the old theory from the perspective of the new theory. (Batterman 2011; Menon and Callender 2013). If the solutions for finite systems *smoothly* approach the solution for infinitely many particles, we have all we need for an explanation of the empirical successes of the idealised theory.

However, it has been argued that in the case of phase transitions the limit is *singular*. As a consequence, the solutions in the limit differ significantly from the neighbouring (i.e., finite-system) solutions, and the latter fail to display phase transitions. Thus, the appeal to infinity appears to be ineliminable for the explanation of the observed phase transitions (Batterman 2011). However, this claim has been disputed (Butterfield 2011; Menon and Callender 2013) on the grounds that the singularity is an artefact of the particular variables chosen (see Hüttemann, Kühn and Terzidis 2015 for discussion).

Furthermore, the following argument can be made in support of the claim that appeal to infinities in the case of phase transitions does not undermine micro-macro coherence: As I mentioned earlier, the behaviour of the compound is described in terms of thermodynamic quantities, and such description involves discontinuous changes in a derivative of a thermodynamic function. But no measurements with finite precision can establish that what we have measured needs to be described as truly discontinuous rather than as something smoother in the vicinity. In other words: As long as our measurements have only finite precision, we can re-describe the observed macro-behaviour as continuous. Even though calculations for systems of finite sizes on the basis of statistical mechanics will never give us phase transitions *as defined in thermodynamics* (namely, as involving discontinuities), they are able to account for empirically indistinguishable behaviour, which characterises what we have observed just as well. This is a version of the fourth option of dealing with threatening cases (see Section 5.5): We can re-describe the macroscopic behaviour to be explained within the limits defined by experimental accuracy. Thus, the case of phase transitions cannot serve as an illustration that we sometimes give up on micro-macro coherence: We do understand how the representations of the compound system and the representation of the parts and their interactions yield the same empirical predictions.

5.6.2 *The Emergence of Classicality*

Another interesting example is what is often called the 'emergence of classicality' (see, e.g., Joos et al. 2003). Classical mechanics describes the behaviour of compound physical systems in terms of definite states; thus, classical systems are typically described as localised. And in many cases, it is empirically adequate to describe macroscopic systems in terms of classical properties. From the perspective of a quantum-mechanical description this is surprising because, unless special circumstances obtain, quantum mechanical systems are expected to be in superposition states and thus not in definite (classical) states.

It seems, therefore, that quantum mechanics cannot explain (or give an account of) why the compound system has definite classical states in terms of the behaviour of the parts of the compound (and their interactions). This might look like a failure of part-whole reduction. So, we may ask: Is the emergence of classicality a case of emergence in the technical sense defined earlier?

There are various suggestions for ways to cope with this issue. I will focus on the decoherence approach. Very briefly, the decoherence approach explains the appearance of a classical property in terms of the interaction of the system in question with the environment (Joos et al. 2003; Schlosshauer 2004). There are two noteworthy features of this approach. First, it is an example of a case that is often encountered in biology and one I have mentioned in the context of protein folding (Section 5.5): In order to avoid the conclusion that the behaviour of the compound system in question is emergent (in the technical sense of Broad, discussed earlier), one can retract the initial assumption that the system under consideration is indeed isolated (option (5) in Section 5.5). Second, what is explained is not why the compound has classical properties given the environment, but rather, why they *appear to be classical to an observer*. This is, thus, also a version of the fourth option of dealing with threatening cases (see Section 5.5): We can re-describe the macroscopic behaviour to be explained (definite classical states) within the limits defined by experimental (or experiential) accuracy.

To conclude: The 'emergence of classicality'[3] may provide an example for the claim that quantum mechanics cannot explain the behaviour of an

---

[3] The notion of emergence that some authors apply in this context and that I used myself, e.g., in the section's heading is not the one used in the technical sense defined in Section 5.5. It is the latter – emergence in the sense of Broad – that is relevant for the question of whether there is a conflict with micro-macro coherence.

isolated macro-system in terms of the behaviour of the constituents and their interactions. However, we still get an explanation of why we can successfully apply macroscopic classical physics to systems with constituents that are characterised in terms of quantum mechanics.

Let me add a remark. In contrast to the example to be discussed in the next section (5.6.3), the case under discussion is one in which the compound and the parts are described in terms of different theories. As a consequence, in this case both theory reduction and part-whole explanation are relevant. An understanding of why the old theory was empirically successful within a certain range and of why we can continue to use it within certain confines is achieved by the decoherence approach as well. Decoherence nicely shows that explaining the past success as well as the continued application of a superseded theory from the perspective of a successor theory does not require deducing the equations of the old theory from the new theory. All that is required is that the solution of the new theory approximates the solution of the old theory (e.g., with respect to localisation) given the appropriate circumstances (being embedded in an environment with which the system interacts).

5.6.3 *Quantum Entanglement*[4]

Quantum entanglement is probably the most promising case for the claim that we have settled for the failure of part-whole explanations. Various authors have considered quantum entanglement as an example of emergence (e.g., Humphreys 1997). Tim Maudlin has even concluded: '[. . .] reductionism is dead. For the total physical state of the joint system cannot be regarded as a consequence of the states of its (spatially separated) parts, where the states of the parts can be specified without reference to the whole' (Maudlin 1998, 54).[5]

While I agree with these claims to some extent, I will try to provide a more nuanced picture. What is important here is a distinction that I briefly alluded to before.

As I mentioned in Section 5.4.2, the explanation of the behaviour of a compound system in terms of the behaviour of the constituents may concern the *state* of a compound system or its *dynamics*. Thus, we might explain why a compound such as an ideal gas has the determinate energy

---

[4] I have discussed this issue in more detail in Hüttemann (2005).

[5] It should be noted that Maudlin when talking about reductionism has in mind primarily Humean supervenience as defined by Lewis. Since I intend to bypass the Humeanism–non-Humeanism debate, I will not discuss how quantum entanglement affects Humean supervenience.

value E* (the macro-state) by pointing out that the constituents have the determinate energy values E^I to E^n (the states of the parts). This would be a part-whole explanation of the *state* of a compound system in terms of the *states* of its constituents. Another question concerns the *dynamics (temporal evolution)* involved. Is it possible to explain the *dynamics* of a compound system in terms of the *dynamics* of the constituent systems?

In the remainder of this section I will argue for the following claims. (1) Quantum entanglement can – at least prima facie – be considered to be a case of emergence (in some sense) – though it is less clear that it is also an example in which we give up on micro-macro coherence. (2) When it comes to the part-whole explanation of the dynamics of compound systems, there is no essential difference between the classical and the quantum mechanical case. They can both be reductively explained.

Let me briefly present quantum entanglement and its implications. Let us suppose we are dealing with a compound quantum mechanical system consisting of two particles. To make things easy we confine ourselves to spin states of the compound system. What are the possible spin states of such a system? (Normalised) vectors in two-dimensional Hilbert spaces represent the spin states of the separate particles. For the construction of a Hilbert space for a compound system, we need a law of composition. According to this law, the possible spin states of the compound system are all those states that can be represented as (normalised) vectors in the tensor product of H_I and H_2, $H_s = H_I \otimes H_2$. H_s contains two kinds of states: Those that can be written as a tensor product of vectors of H_I and H_2 respectively and vectors that cannot be expressed in such a way.

If we take as a basis for H_I the eigenvectors in the spin z-direction $|\psi^{z\text{-}up_I}\rangle$ and $|\psi^{z\text{-}down_I}\rangle$ and as a basis for H_2 $|\psi^{z\text{-}up_2}\rangle$ and $|\psi^{z\text{-}down_2}\rangle$, we find all of the following among the possible states of the compound system (normalised):

(i) $|\psi^{z\text{-}up_I}\rangle \otimes |\psi^{z\text{-}down_2}\rangle$

(ii) $|\psi^{z\text{-}up_I}\rangle \otimes |\psi^{z\text{-}up_2}\rangle$

(iii) $1/\sqrt{2}\,|\psi^{z\text{-}up_I}\rangle \otimes |\psi^{z\text{-}down_2}\rangle - 1/\sqrt{2}\,|\psi^{z\text{-}down_I}\rangle \otimes |\psi^{z\text{-}up_2}\rangle$

The essential point is that (iii) cannot be written as a simple tensor product of vectors of H_I and H_2. It can only be written as a superposition of such tensor products. In terms of physics, this amounts to the following. There are spin states of the compound system such as (iii) that do not allow the attribution of pure states to the parts of the compound. So, the fact that the compound is in a determinate state cannot be explained in terms of determinate states the constituents occupy. There is no part-whole explanation because there are no pure states of the parts to start with.

This is the kind of emergence that Paul Humphreys discussed in Humphreys (1997). He starts with a notion he calls 'fusion', which he defines as follows:

> The key feature of [a fused system] is that it is a unified whole in the sense that its causal effects cannot be correctly represented in terms of the separate causal effects of [the components]. Moreover, within the [fused system] the original property instances [. . .] no longer exist as separate entities and they no longer have their [. . .] causal powers [. . .]. (Humphreys 1997, 10)

The essential feature of this example in terms of Humphreys is what he calls the failure of supervenience: The state of the compound cannot supervene on (let alone be explained in terms of) the states of the parts because there are no such states. Humphreys then suggests that the failure of supervenience implies emergence (Humphreys 1997, 10).

Humphreys' notion of emergence differs from Broad's notion. For Broad, the supervenience of the behaviour (or states) of the compound on that of the parts is a condition for emergence. For Broad, given supervenience, the failure of adequately explaining the state of the compound by relying on that of the parts constitutes emergence. This is important for our discussion because it indicates that it is not quite clear that quantum entanglement does indeed provide an example in which we have settled for not being able to explain the state of the compound in terms of the states of the parts, because there are no states of the parts in an entangled state to start with (according to Humphreys). So, while I agree that from a classical perspective it is deeply disturbing that there are systems formed by 'fusing', say, two electrons such that we are no longer able to attribute definite states to the individual electrons, it is nevertheless not the case that we have a situation in which we have the state of the compound and the states of the parts and we cannot explain the former in terms of the latter. A precondition for this sort of failure is that the parts have definite states.

Maybe this argument is not completely convincing. The worry is, after all, that we cannot give a part-whole explanation of the state of the compound even though there are parts (if we assume there are parts, which seems to me to be a reasonable assumption in the light of what we will say about the dynamics of the system). Presumably, I should agree that this case is in tension with the regulative ideal of micro-macro coherence. Still, the reason for this tension is not that what we know about the state of the parts and what we know about the state of the compound does not fit but, rather, that there are no states for the parts to start with. Thus, even though some of our expectations have been frustrated, what we called 'micro-macro coherence', i.e., the requirement that the representations of the compound system and the representation of the parts and their interactions should yield the same predictions with respect to the behaviour of the system in question, has in the literal sense not been violated.

Let me now turn to the part-whole explanation of the *dynamics* of a compound quantum mechanical system. In quantum mechanics, a vector in Hilbert space represents the state of a physical system at a time t. The time-evolution of this state-vector (the dynamics) is described by the Schrödinger-equation, which in turn requires the quantum mechanical Hamiltonian to be specified.

In the case of an isolated one-particle system, the quantum mechanical Hamiltonian is defined as $H = \frac{1}{2} P^2/m$, where P is the momentum operator of the particle. When we turn to a system of two non-interacting particles the procedure is analogous to the classical case. We rely on a quantum mechanical law of composition. It essentially requires that we take the tensor product of the two Hilbert spaces so as to gain a new Hilbert space in which the two-particle system can be represented. As a consequence, the Hamiltonian for the combined system is a direct sum of those for the isolated subsystems (see, e.g., Bohm 1986, 147 or Kennedy 1995, 543–60).

This presentation of part-whole explanation in quantum mechanics suggests that not much has changed while turning from classical mechanics to quantum mechanics. That is indeed what I maintain – as long as we confine ourselves to the part-whole explanation of the *dynamics* of compound systems. The laws that govern the temporal evolution of compound quantum mechanical systems can be explained in terms of those laws that would govern the isolated components plus the laws of composition (plus interaction laws).

Another example is the ideal crystal that we discussed in Section 5.4.2. Hardly anything changes with respect to the explanatory strategy when we go from classical to quantum mechanics. The point is that though the mathematical tools used to describe the systems and subsystems change, the reductive analysis of the ideal crystal remains the same. The quantum mechanical Hamilton operator for the ideal crystal is determined on the basis of the kinetic energy terms for the ions and their interactions. The variables p and q are replaced by operators on Hilbert space. This replacement eventually leads to different empirical predictions (e.g., with respect to the temperature dependence of the specific heat). However, there is no change with regard to the explanatory strategy that is involved.

The fact that the *states* of quantum mechanical systems cannot, in general, be explained in terms of the *states* of their constituents does not undermine part-whole explanation in general. The explanation of the behaviour of a crystal is an example for the explanation of the *dynamics* of the system. The dynamics of quantum mechanical systems is essentially determined by the Hamiltonian that goes into the Schrödinger equation (or the partition function). The Hamiltonian is built up by what we know about how the parts would evolve if they were separated and their interaction while constituting a compound; it is built up of the terms for the kinetic and potential energies. This information about the parts and their interaction is sufficient to explain the dynamics of the compound system. The upshot is that the quantum mechanical explanation of the *dynamics* of compound quantum systems is just as reductionist as its classical analogue.

Even though quantum entanglement does provide a challenge for the claim that aiming for part-whole explanation is a regulative principle for scientific research, it would be premature to claim that micro-macro coherence no longer plays the role of an epistemic constraint we are committed to.

5.7 Ontological Implications of Reductive Practices

In the previous sections we have looked at reductive practices in the sciences. I have argued that we are interested in pursuing these practices because we are interested in achieving an understanding of how different descriptions of either one system (part-whole explanations) or classes of systems (theory reduction) are related.

Theory reduction is motivated by the requirement that the empirical success of the old theory as well as its continued application needs to be accounted for from the vantage point of the new theory. One way of achieving this is by Nagel reducing the old theory to the new theory. Most likely there are not many examples of Nagel reduction. Limit-case reduction as discussed in Section 5.3 would suffice, but there may be further ways of accounting for the empirical success of the old theory. Decoherence theory – among other things – explains what has been called the 'emergence of classicality' and thus contributes to an understanding of why classical mechanics is empirically successful in certain domains without exemplifying either Nagel reduction or limiting-case reduction.

What I called 'micro-macro coherence' is relevant when we have two descriptions of one system, typically one of which is a (more) macroscopic description whereas the other is a (more) microscopic description: The representations of the compound system and the representation of the parts and their interactions should (i) yield the same predictions with respect to the behaviour of the system in question and (ii) account for why they yield the same predictions in terms of general laws. For example, we want to understand how an intervention on a part affects the compound and, conversely, how an intervention on the compound affects the parts. For such an understanding, we need a detailed general account of how what we know about the parts is connected with what we know about the compound, i.e., we need a part-whole explanation.

Aiming for theory reduction and aiming for part-whole explanation are built into scientific practice as regulative principles. *There is thus no need to assume Foundationalism or Eliminativism* or any other metaphysical hypothesis to make sense of why we are interested in these practices. The rationale for both theory reduction and part-whole explanation can be spelt out purely in terms of epistemic or pragmatic rather than ontological commitments.

There are various possible objections to this line of argument. It might, for instance, be argued that our account only partially answers why we are interested in this reductive practice. It can be agreed that *once we have descriptions of the compound and of its parts* then it might very well be a regulative principle of scientific practice that we want these descriptions to fit together. While this is an interesting observation, there is another issue that I haven't addressed: Why are we interested in the behaviour and interaction of the constituents or parts to start with? It is one thing to ask for a rationale for why, *given* descriptions of the compound and descriptions of the parts, these have to fit together. It is quite another thing to explain why we go micro in the first place. Why is it that we are interested in exploring the constitution of systems? Is it here, perhaps, that Foundationalism or Eliminativism needs to be presupposed?

I think there is an obvious way to see that neither Foundationalism nor Eliminativism is needed to account for why we (sometimes) go micro. The analogy with causes is instructive: Why are we interested in the type-level causes of the behaviour of systems? There is no need to postulate that causes are more fundamental than effects in order to give a rationale for our interest in causes. We are interested in causes to a large extent because they allow us to manipulate systems. Causes give us handles to bring about the changes we are interested in. Causal explanations trace the dependence relations that are required for such manipulations. Similarly, I would argue, we can give a pragmatic account for our interest in constitutional relations: They give us additional handles to manipulate the behaviour of systems. Having a grip on the behaviour of the parts of systems gives us fine-grained handles to bring about the changes we wish to make. While causation allows diachronic manipulation, part-whole relations allow synchronic manipulation. So, the short answer is that we go micro because this increases our options to manipulate systems. Part-whole explanations trace the dependence relations that are required for such manipulations. There is no need for metaphysical postulates in accounting for why we are interested in the parts of compound systems.

Another objection is that the explanatory practice is asymmetric but micro-macro coherence is not. Where does this asymmetry come from? Does it require an ontological asymmetry? Furthermore, it might be argued that mereological relations are ontological dependence relations and that, therefore, ontological assumptions do play a role in accounting for our practice of part-whole explanations. These and similar questions do not concern the *rationale* for our reductive practice but, rather, the question of whether the best explanation of the success of these practices requires certain metaphysical assumptions. These issues will be dealt with in the next chapter. The point of the present chapter – as mentioned at the outset – was, first, to characterise our reductive practices and, second, to establish that there is an epistemic/pragmatic account of the rationale for these practices, i.e., of why we are interested in these reductive practices.

CHAPTER 6

Reduction and Physical Foundationalism

In the last chapter, I looked at some reductive explanatory practices and argued that we can explain why we pursue these reductive practices by appeal to our interest in achieving an understanding of how different descriptions of either one system (part-whole explanations) or of classes of systems (theory reduction) are related.

In the case of a compound system this quest explains why, given an account of the behaviour of the parts and another account of the behaviour of the compound, we want these accounts to fit together. Foundationalism (the view that all other facts, events or entities are ontologically posterior vis-à-vis physical facts, events or entities) or other ontological assumptions are not needed as part of such an explanation. I furthermore argued that there are pragmatic reasons for why we are interested in investigating the behaviour of the parts: A detailed knowledge of their behaviour and their interactions allows a fine-grained manipulation of the behaviour of the compound. Neither our interest in investigating the micro-level nor our interest in part-whole reduction requires ontological assumptions.

Nevertheless, even though we have given a purely epistemic/pragmatic account of the rationale for our reductive practices, an explanation of why these practices work might commit us to ontological assumptions that eventually imply either Foundationalism or Physical Eliminativism. It is the purpose of this chapter to examine to what extent the success of the reductive practices commits us to Foundationalism. Chapter 7 will be concerned with Physical Eliminativism.

I will argue that neither Foundationalism nor Physical Eliminativism is implied by our explanatory practices. I am not denying that Foundationalism might consistently be taken to be true. I will, however, argue that Foundationalism does not do any work in accounting for the success of scientific practice. A minimal metaphysics, which only accepts metaphysical claims that contribute to explaining scientific practice, should thus abstain from Foundationalism.

6.1   Foundationalism

At the outset of Chapter 5, I hinted at the claim that our reductive practices might suggest the view that fundamental physics gives us *the real story* of what is going on in the world.

Such a claim, I argued, might be spelled out in different ways. It might be read as an eliminativist claim: It is only the fundamental physical facts or objects that exist; macroscopic everyday facts or objects only appear to exist. The view that fundamental physics gives us *the real story* of what is going on in the world might also be construed as a foundationalist claim about ontological priority or metaphysical dependence. According to this reading, it is the view that all other facts or objects obtain *in virtue of* or are *metaphysically dependent on* more fundamental physical facts or objects. The latter facts or objects are *ontologically prior* to all other facts or objects. I will now make the idea of Foundationalism as precise as we need it to be for the purposes of this chapter.

In the last 25 years or so, the idea that certain dependence relations need to be spelt out in non-modal terms has gained some traction. Kit Fine, for instance, argued that the essence of an entity should not be equated with modal relations:

> Consider, then, Socrates and the set whose sole member is Socrates. It is then necessary, according to standard views within modal set theory, that Socrates belongs to singleton Socrates if he exists; for, necessarily, the singleton exists if Socrates exists and, necessarily, Socrates belongs to single-ton Socrates if both Socrates and the singleton exist. It therefore follows according to the modal criterion that Socrates essentially belongs to single-ton Socrates. But, intuitively, this is not so. It is no part of the essence of Socrates to belong to the singleton. (Fine 1994, 4–5)

Similarly, Jonathan Schaffer argued that some sort of non-modal dependence relation is needed to account for the dependence relations that obtain in our world ('non-modal dependence relation' here means that the dependence relation in question cannot be explicated completely in modal terms):

> The mereological structure of whole and part is not the only structure to the world. There is also the metaphysical structure of prior and posterior, reflecting what depends on what, and revealing what are the fundamental independent entities that serve as the ground of being. Consider Socrates. Given that he exists, the proposition <Socrates exists> must be true. And conversely, given that the proposition <Socrates exists> is true, there must be Socrates. Yet clearly there is an asymmetry. The proposition is true

because the man exists and not vice versa. Truth depends on being. (Schaffer 2010, 35)

There is now a considerable literature discussing various kinds of non-modal dependence relations, e.g., under the heading of 'grounding' or 'ontological dependence'.[1] What is relevant for the purposes of this paper is that Metaphysical Foundationalism is a view that is usually spelt out in terms of such non-modal dependence relations. Ricki Bliss in a recent paper sketches this view as follows:

> Proponents of the kind of view I have in mind will claim that reality is hierarchically arranged by chains of entities ordered by relations of ground [i.e. by a non-modal dependence relation, Author], where that relation is asymmetric, irreflexive, transitive and well-founded. This view has come to be known as *metaphysical foundationalism*, and it is widely thought to be at least true of the actual world. (Bliss 2019, 361)

For the purposes of this chapter, I will use the following terminology:

> Metaphysical Foundationalism obtains if the following conditions are satisfied:
>
> (1) Ontological Priority: Different kinds of facts/entities/events can be ordered such that some are ontologically more fundamental than others (they are ontologically prior to others).[2]
> (2) Well-foundedness: There is a set of facts/entities/events that is fundamental, i.e., a set of facts/entities/events that is ontologically prior to all other facts, such that every fact/entity/event either is grounded in some other fact/entity/event or is fundamental.

Some remarks: First, I will use the term 'ontological priority' to describe the non-modal dependence relation in question to indicate that I am only interested in *asymmetric* ontological dependence relations. Second, there is a debate about whether ontological dependence relations should be reconstructed in terms of sentential operators or relations (see, e.g., Fine 2012, 43). I will conceive of ontological priority as a relation. Third, there is the question of the relata of this relation. Sometimes these relata are taken to be entities (e.g., Schaffer 2010), sometimes facts (e.g., Rosen 2010) and

---

[1] There is also a considerable amount of scepticism about the grounding relation (see, e.g., Wilson 2014 or Hofweber 2009). I introduce Foundationalism and the non-modal dependence relation that comes with it as a hypothesis (for the explanation of the success of certain aspects of scientific practice) only, which will be ultimately rejected. There is thus no need for me to get into the details of this discussion.

[2] The question of whether this ordering is full or partial will play no role in what follows.

sometimes events (e.g., Kim 1994 [2010]). In this book, I assume that I do not need to commit myself to one of these views. Fourth, in debates about Metaphysical Foundationalism it is usually (2), i.e., well-foundedness, that is at issue (see, e.g., Morganti 2015). By contrast, my focus in what follows will be on (1), i.e., on ontological priority. Finally, Metaphysical Foundationalism is silent about the nature of the fundamental facts/ entities/events. In the context of our discussion the natural candidates should be physical facts, etc. 'Physical Foundationalism' would then be an appropriate name. However, because the nature of the fundamental facts/entities/events plays no role in the remainder of the chapter, I will simply use the term 'Foundationalism'.

For the purposes of this chapter, I only need to appeal to two features of ontological priority that – given ontological priority is meant to be constitutive for Foundationalism – I take to be necessary conditions for ontological priority:

(i) *Asymmetry*: If a fact/entity/event F1 is ontologically prior to another fact/entity/event F2, then it is not the case that F2 is ontologically prior to F1.

Asymmetry is a necessary condition for ontological priority if the latter notion is taken to be able to articulate Foundationalism:

> one key task for the notion of fundamentality [i.e. Foundationalism] is to help us articulate the view that there is a *hierarchical structure* to reality. (Tahko 2018, original emphasis)

A symmetrical dependence relation (as, e.g., advocated by Barnes 2018) would be of no use to articulate the hierarchical structure that goes with Foundationalism.

(ii) *Non-modality*: A purely modal conception of ontological priority would be unable to account for some well-known intuitive cases: Necessarily, if Socrates exists the empty set exists, but not vice versa. So, we have an asymmetric necessitation relation. However, it is fairly plausible that Socrates is not ontologically dependent/posterior relative to the empty set. Asymmetric necessity is thus not sufficient to capture ontological priority. (Let me stress that when I use the term 'non-modal' in this chapter, it is used as in the grounding literature – a non-modal relation is hyperintensional relation. This contrasts with the use in the Humeanism debate, where a 'non-modal' relation is a relation that is neither modal nor hyperintensional.)

In Chapters 1 and 2, we have allowed ourselves a certain inventory of metaphysics: Systems having multi-track dispositions, which underlie laws of nature. We also postulated that these laws hold with some sort of necessity that can be spelled out in terms of invariance relations, where invariance is a modal notion. If Foundationalism is true, we need something additional over and above the nomological connections or invariance relations we have already introduced. We need relations that have further features: They need to be asymmetric and they need to be non-modal, i.e., not completely explicable in modal terms (though, presumably, implying modal relations).

The pattern of argument in the following sections will be that I will look at various features of the reductive practices discussed in Chapter 5 and examine for each of these whether it commits us to Foundationalism. More precisely, I will examine whether we can account for the success of the various features of the reductive practices by appeal to the metaphysical inventory of Chapters 1 and 2 or whether we additionally need to appeal to an asymmetric and non-modal dependence relation. If the reductive practices do not commit us to the claim that an ontological priority relation obtains in nature, Foundationalism will not be implied and a Minimal Metaphysics of Scientific Practice will thus refrain from Foundationalism and ontological priority.

6.2   Composition and Ontological Priority

Explaining the behaviour of compound systems in terms of their parts may mean more than one thing. Healey (2013) has distinguished various senses of part-whole relations and explanations within physics. He has also pointed out that one and the same system might be decomposed in various ways even if the sense of decomposition is fixed. In what follows I will bypass these problems and will grant that there are at least some bona fide part-whole explanations. I will focus on those part-whole explanations that seem to be the most promising for establishing claims of ontological priority. Not even in these cases, I will argue, will the successful explanatory practice warrant Foundationalism.

The practice of part-whole explanations, it may be argued, presupposes that there are compounds and that these have parts. The relation of parts and wholes is a dependence relation, so, naturally, the question arises whether this implies that in part-whole explanations we are committed to ontological priority.

Physical compounds or wholes like Theseus' ship do not depend for their existence on the existence of particular parts (e.g., planks). However,

Theseus' ship does depend for its existence on there being planks. One way of explicating these claims is by distinguishing:

- *Rigid existential dependence:* x depends for its existence upon y = $_{df}$ Necessarily, x exists only if y exists.
- *Generic existential dependence:* x depends for its existence upon Fs = $_{df}$ Necessarily, x exists only if some Fs exists. (Tahko 2015, 96)

Theseus' ship – or any other compound – does not depend for its existence on the existence of a particular plank, so rigid existential dependence plays no role. It does, however, depend for its existence on there being planks at all, so generic existential dependence obtains between a compound and its parts. Furthermore, this dependence is asymmetric. While the existence of certain sorts of entities (parts) is necessary for the existence of a particular compound, the converse is not true.

So, assuming that part-whole explanations entail that a compositional relation obtains between the entities referred to in the explanans and those mentioned in the explanandum, we are committed to the obtaining of an asymmetric dependence relation between kinds of entities that serve as parts and the compound with respect to existence.

As we will see, this is the closest we will come to establishing the relation of ontological priority, because there is at least a certain sort of asymmetry here. However, asymmetric generic existential dependence is a purely modal concept (it can be defined in terms of necessitation) and thus not apt to capture the non-modal dependence required for ontological priority (Fine 1995; Correia 2008, 1024).

The following worry might be raised: People do take generic existential dependence to be stronger than merely modal because it is explanatory. It is this 'explanatoriness' that requires a non-modal relation. I will look at this argument and the 'explanatoriness' of the physical part-whole relation in the next section.

To summarise the discussion up to this point: Our reductive practices might commit us to composition, but this on its own does not commit us to Foundationalism because it does not commit us to the obtaining of an ontological priority relation between parts and wholes, where the latter would be a non-modal relation.

6.3   Part-Whole Explanation and Ontological Priority

Let us now turn to the 'explanatoriness' of physical composition. How possibly could the fact that we can explain the behaviour of a compound system in terms of the behaviour of the parts, their interactions, etc. commit us to the obtaining of an ontological priority relation and thus, possibly, to Foundationalism?

The key to establishing a connection is explanatory realism, a view articulated by Jaegwon Kim:

> What I want to call *explanatory realism* takes the following form: C is an explanans of E *in virtue of* the fact that c bears to e some determinate relation R. Let us call R, whatever it is, an 'explanatory relation'. (The *explanans* relation relates propositions or statements. The explanatory relation relates events or facts in the world.) (Kim 1988 [2010], 149, original emphasis).

Elsewhere Kim uses the slogan '*explanations track dependence relations*' (Kim 1994 [2010], 184). The dependence relations in question are also known as *metaphysical backing relations* (Ruben 1990, 2). According to explanatory realism, part of what is necessary for a successful explanation is that the facts, events, etc. referred to in the explanans determine or fix the facts, events, etc. referred to in the explanandum. In a causal explanation, it is the causal relation that plays the role of a backing relation for the explanation: For a causal explanation to succeed, the cause has to (at least partially) determine the effect – the effect has to depend on the cause cited. The causal explanation tracks the causal dependence relation.

In what follows, I will assume that Kim is right: Explanations commit us to dependence relations. The relevant question for our purposes is whether we need to assume that the backing relation that is tracked in physical part-whole explanations is asymmetric and non-modal. My main focus will be on asymmetry. We can formulate a two-step argument to the effect that the availability of part-whole explanations commits us to the assumption that an asymmetric backing relation obtains between parts and wholes. The first step establishes that part-whole explanations commit us to assuming dependence relations between parts and wholes.

(1) *Part-whole explanation*: Physics provides part-whole explanations.
(2) *Explanatory realism*: Explanations track dependence relations.
(3) Conclusion 1: Part-whole explanations track a dependence relation (backing relation) (from (1) and (2)).

In the second step, it is argued that this dependence relation is asymmetric.

(4) *Explanatory asymmetry*: We explain the behaviour of compound physical systems (wholes) in terms of the behaviour of their parts, but not vice versa.

(5) *Inheritance*: Explanatory relations inherit their formal features from the relevant backing relation.

(6) Conclusion 2: The backing relation for part-whole explanations is asymmetric (from (4) and (5)).

This two-step argument might be taken to be part of a more comprehensive argument for the claim that what happens at the macro-level happens *in virtue* of what happens at the micro-level and for the claim that the behaviour of the parts is ontologically prior relative to the behaviour of the compound. As already indicated, in what follows I will be mainly concerned with the question of whether we are committed to an asymmetric backing relation.

I have no objections to the first step of the argument. The second step of the argument, even though valid, isn't sound. *Inheritance*, i.e., premise (5), is the culprit. Explanatory realism commits us to (2) but not to (5).[3] Consider the following case. Suppose we are dealing with an electronic network, i.e., a collection of interconnected electronic components. What we are interested in is the manipulation of voltages across and currents through these components. A change in voltage will lead to a change in current. A change in current will lead to a change in voltage. Suppose we want to explain why a particular change in voltage ΔV leads to a change ΔI in current. The explanation will presumably rely on Ohm's law: $V = RI$, where R is the resistance of the material in question. The explanation is asymmetric from voltage to current. The dependence of the current on the voltage is described by Ohm's law. But Ohm's law does not describe an *asymmetric* dependence relation between voltage and current. Ohm's law is completely symmetric in this respect. We can use it just as well to explain changes in voltage due to manipulation of current. The asymmetry of the explanation is not due to an underlying asymmetric dependence relation but, rather, due to the fact that we are interested in manipulating one quantity rather than the other. What the example illustrates is that an asymmetry in an explanation does not commit us to the claim that the underlying dependence relation (backing relation) is asymmetric. *Inheritance* ought to be rejected.

Another example is causal explanation. Causal explanation is asymmetric. This asymmetry does not necessarily commit us to the view that the underlying laws in virtue of which the effect depends on the cause have to

---

[3] Kim, however, seems to take it for granted that backing relations are asymmetric (Kim 1994 [2010], 184).

be asymmetric. The fundamental laws of physics might very well be time-symmetric. The asymmetry of causal explanations might be due to the initial state of the universe or to pragmatic features of explanations. This is, of course, a controversial issue. But, it is by no means clear that the asymmetry of causal explanation is incompatible with an underlying symmetric dependence relation that serves as the backing relation for causal explanations.

In the remainder of this section, I will have a closer look at physical part-whole explanations and at the backing relation involved in such explanations. In this context, it is important to keep two issues separate. The first concerns the question of how a compound system with *respect to its existence* depends on the *existence* of (kinds of) parts. That is what we dealt with in Section 6.2. This has to be kept apart from the following question:

> Given compound systems with their parts (and given that asymmetric generic existential dependence holds between the compound and its parts), how does the behaviour of the compound depend on the behaviour of the parts? What kind of dependence relation obtains between the behaviour of the compound and the behaviour of the parts – once the compound exists?

It is this latter dependence relation that we turn to when we ask for ontological implications of part-whole explanations. I will argue that physics describes the dependence of the behaviour of compounds on that of the parts in terms of a dependence relation that is mutual and thus not asymmetric.[4] The asymmetry of the part-whole explanation should thus not be attributed to an asymmetric dependence or backing relation.

Let us start with the notion of *behaviour*. With respect to the behaviour of a physical system, we can distinguish the state of the system, its constants and its temporal evolution. Some quantities of a physical system are constant; others vary with time. In the case of classical particles, we can, for instance, distinguish their positions and momenta as changing quantities, while other quantities that might be relevant for the system under consideration, such as the gravitational constant, remain constant. The values of the variable quantities at a particular time are called the state of the physical system at this time. However, the constants and the state of a system at a certain time do not exhaust what is commonly understood as the system's behaviour. In addition, we have laws that describe the

---

[4] I have argued for this claim in more detail in Hüttemann (2004, 2015).

connections between the various quantities involved; in particular, the way the state of the system develops in time. What these laws describe is the temporal evolution or dynamics of the system. Explaining the behaviour of compound systems in terms of their parts may refer either to the state or to the dynamics.

Part-whole explanation of the *state* of a compound system explains the state at a certain time in terms of the states of the parts at that time. Thus, we might explain why a compound system, such as an ideal gas, has the determinate energy value E^* (the macro-state) by pointing out that the constituents have the determinate energy values E^1 to E^n (the states of the parts).

Quantum entanglement is a prominent *counter*example to this kind of part-whole explanation, if the *explanans* is restrained to *actual* states of the system. It is not, in general, possible to explain the state of compound quantum mechanical systems in terms of actual states of the parts because quantum mechanics does not, in general, specify states for the parts where the states of the parts can be specified without reference to the whole (see, e.g., Maudlin 1998).

This might be taken to be bad news for the foundationalist (assuming that the evidence for Foundationalism relies on successful part-whole explanations), but not as bad as it might seem. First, as we have seen in Chapter 5, this failure need not be taken as a counterexample to micro-macro coherence in general. More important in the current context is the fact that there is another dimension to part-whole explanation, namely, part-whole explanation of the dynamics of the compound system, which is not confronted with counterexamples from quantum mechanics (see Section 5.6.3 as well as Hüttemann 2005 for the distinction between part-whole explanation of states and part-whole explanation of the dynamics).

Part-whole explanation of the dynamics of a compound specifies the temporal evolution or dynamics of the system in terms of the dynamics of the parts (plus interactions among the parts). In what follows, I will focus exclusively on the part-whole explanation of the dynamics of a system because it provides the best starting point to argue that by part-whole explaining we are committed to Foundationalism.

So, how does this kind of part-whole explanation work? For purposes of illustration I will start with a simple example: a non-interacting two-particle system. The first step in the explanation or analysis of the dynamics of this system is the identification of its parts, i.e., the two (isolated) one-particle systems.

The second step consists in determining the dynamics of the isolated one-particle system. According to classical mechanics, the complete behaviour of a one-particle system is specified by its path in six-dimensional phase space. A point in phase space represents a state of a classical system. The Hamilton equations specify the system's time-evolution or dynamics and thus its path in phase space. These equations in turn require a classical Hamilton function. The dynamics of an isolated particle, for instance, can be described by a classical Hamilton function of the form $H = p^2/2\,m$, where p is the momentum and m the mass of the isolated particle.

For a non-interacting two-particle system we first need to specify two six-dimensional phase spaces, one for each of the particles, as well as a classical Hamilton function of the form in the previous paragraph for each of them. That, however, is not yet a description of a two-particle system. It is a description of two separate one-particle systems.

What we need in addition is something that tells us how the descriptions of the behaviour of subsystems have to be combined so as to obtain the description of the behaviour of the compound system. We basically need the following information. 1) The phase space for a compound system is the direct sum of the phase spaces of the subsystems. Thus, for the two-particle system we obtain a 12-dimensional phase space. 2) The Hamilton function for the compound system is the sum of those for the isolated constituents. Thus, the dynamics of the system of two non-interacting particles in classical mechanics is described by a Hamilton function of the form $H = p_1^2/2m_1 + p_2^2/2\,m_2$.

This is the third and final step in the explanation or analysis of the dynamics of the non-interacting two-particle system: adding up the contributions of the parts according to laws of composition.

If interactions are present, then we have to introduce a further term into the Hamiltonian, e.g., a term for gravitational interaction such as $-Gm_1\,m_2/r$, where G is the gravitational constant and r the distance between the two particles.

Let me come back to an example from quantum mechanics that I have already discussed in Section 2.4.3. Carbon monoxide molecules consist of two atoms of mass m_1 and m_2 at a distance x. The compound's (the molecule's) behaviour is explained in terms of the behaviour of two subsystems, the oscillator and the rotator.[5]

---

[5] The fact that these parts are not spatial parts but, rather, sets of degrees of freedom is irrelevant for our discussion – if they were spatial parts, the structure of the explanation would not be different, as the classical example, discussed earlier, illustrates. What is essential is that physical part-whole explanations rely on laws of composition – whether or not the parts are spatial. If there is a problem at all, it is a problem for those who want to argue from physical part-whole explanations to Foundationalism. If the part-whole relation as described in physics does not properly map onto the part-whole relation as described in metaphysics, that would make it even more difficult to argue that part-whole explanations commit us to Foundationalism.

The first step of the part-whole explanation of the dynamics of the compound system consists in the identification of its parts. In a second step, the dynamics of the isolated subsystems is determined. In particular, Bohm considers the following subsystems: (1) a rotator, which can be described by the Schrödinger equation with the Hamiltonian $H_{rot} = L^2/2I$, where L is the angular momentum operator and I the moment of inertia; (2) an oscillator, which can be described by the Schrödinger equation with the following Hamiltonian: $H_{osc} = P^2/2\mu + \mu\omega^2 Q^2/2$, where P is the momentum operator, Q the position operator, ω the frequency of the oscillating entity and μ the reduced mass.

It is essential that laws concerning constituents considered in isolation are never sufficient to explain even the simplest kinds of compound systems. We always need a law of composition that tells us how exactly the behaviour of the subsystems determines that of the compound, e.g., in terms of prescriptions of how to construct a Hilbert space for the compound system given the subsystems. The third step thus consists in adding up the contributions of the subsystems. This is done by invoking a law of composition, which I have already quoted in Section 2.4.3:

> IVa. Let one physical system be described by an algebra of operators, A_1, in the space R_1, and the other physical system by an algebra A_2 in R_2. The direct-product space $R_1 \otimes R_2$ is then the space of physical states of the physical combinations of these two systems, and its observables are operators in the direct-product space. The particular observables of the first system alone are given by $A_1 \otimes I$, and the observables of the second system alone are given by $I \otimes A_2$ (I = identity operator). (Bohm 1986, 147)

The explanatory strategy in this example as well as in the classical example discussed earlier can be summarised as follows: The behaviour of the compound system (the explanandum) is explained in terms of the behaviour of the parts (the explanans) by relying on a law of composition. The law of composition describes how what is referred to in the explanandum depends on what is referred to in the explanans.

So, if we are interested in whether the dependence or backing relation for part-whole explanations is asymmetric, we should have a closer look at the law of composition.

The law of composition for quantum mechanics gives us a prescription for the Hamiltonian that describes the behaviour of a compound system. In the absence of interactions, we have, strictly speaking, the following:

$$H_{comp} = H_1 \otimes I_2 \otimes I_3 \ldots \otimes I_n + I_1 \otimes H_2 \otimes I_3 \ldots \otimes I_n + \ldots I_1 \otimes I_2 \otimes I_3 \ldots \otimes H_n$$

The index i ranges over all subsystems and I_n is the identity operator for the n-th subsystem's Hilbert space. That looks somewhat cumbersome. Instead, we typically encounter the considerably simpler

$$H_{comp} = H_1 + H_2 + \ldots H_n$$

Let us consider the case of a compound consisting of three subsystems. Thus, we have

$$H_{comp} = H_1 + H_2 + H_3$$

The law of composition ensures that the behaviour (dynamics) of the subsystems (H_1, H_2 and H_3, respectively) determines the behaviour (dynamics) of the compound (H_{comp}).

The determination relation is represented by an equation. Once the three Hamiltonians on the right-hand side are specified, so is the fourth for the compound on the left-hand side. But obviously, the same is true for any of the other Hamiltonians as well. If H_{comp}, H_1 and H_2 are given, H_3 is determined according to the equation

$$H_3 = H_{comp} - H_1 - H_2$$

Each of the four is determined as soon as the other three are fixed. The relation between the subsystems and the compound with respect to determination of behaviour is *mutual*:

> In other words: The backing relation fails to be an asymmetric dependence relation.

For the dependence relation as described by the law of composition, both the following hold:

- The compound's behaviour depends on the behaviour of part 3, because the compound's behaviour is partially determined by that of the third component or part.
- Part 3's behaviour depends on the behaviour of the compound, because the behaviour of the third component is partially determined by that of the compound.

Since the compound's behaviour depends on the behaviour of part 3 and vice versa, the dependence relation, as described by the laws of composition, fails to be asymmetric.

One may object that the non-interaction case is rather trivial and not very interesting. Taking interactions into account does indeed complicate the picture. But the complications have to do with the question of what to consider as the parts in a part-whole explanation with interactions rather than with the *nature* of the relation between the behaviour of parts and the behaviour of wholes. When a foundationalist argues that part-whole explanations commit us to the view that the behaviour of the parts is ontologically prior relative to that of the compound, she has to specify what she means by 'the behaviour of the parts'. I will consider two specifications and argue that in both cases, the same conclusions hold as in the non-interaction case.

Let us take a classical case with interaction. In the presence of interactions, we have to introduce a further term into the Hamiltonian, e.g., a term for gravitational interaction such as $-Gm_1 m_2/r$, where G is the gravitational constant and r the distance between the two particles. In such a case, the foundationalist probably has two options of describing what an explanation in terms of the behaviour of the subsystems might mean. According to the first (very natural) option, the relevant subsystems are the isolated particles in the absence of any forces acting on them. In order to explain the compound's behaviour, we do not only rely on the general law of composition; the term for the gravitational field potential has to be added as well. This reading of 'the behaviour of the parts' accords with the claim that the compound's behaviour is explained in terms of the behaviour of the parts *and their interactions*. This yields the following Hamilton function for the compound system:

$$H_{1+2} = p_1^2/2m_1 + p_2^2/2\, m_2 - Gm_1\, m_2/r$$

or

$$H_{1+2} = H_1 + H_2 - Gm_1\, m_2/r$$

Again, the determination relation holds because we are dealing with an equation, and once the three terms on the right-hand side are specified, so is the fourth for the compound on the left-hand side. But, as before, the same is true for any of the other terms as well. If H_{1+2}, $- Gm_1\, m_2/r$ and H_2 are given, H_1 is determined according to the equation $H_1 = H_{1+2} - H_2 + Gm_1\, m_2/r$.

Each of the four terms is determined as soon as the other three are fixed. The relation between the behaviour of the subsystems, the interaction and the behaviour of the compound with respect to dependence is *mutual*.

We get the same conclusion as in the non-interaction case: Both of the following claims come out as true:

- The compound's behaviour depends on the behaviour of part 2, because the compound's behaviour is partially determined by that of the third component or part.

- Part 2's behaviour depends on the behaviour of the compound, because the behaviour of the third component is partially determined by that of the compound.

The law of composition thus specifies how the behaviour of the parts plus the interaction determines the behaviour of the compound. What we have here again is that according to the law of composition, a dependence relation holds, and once the three terms on the right-hand side are specified, so is the fourth for the compound on the left-hand side. However, the same is true for any of the other terms as well. If H_{1+2}, $Gm_1 m_2/r$ and H_2 are given, H_1 is determined according to the equation $H_{1+2} = H_1 + H_2 - Gm_1 m_2/r$. Each of the four terms is determined as soon as the other three are fixed. The relation between the behaviour of the subsystems, the interaction and the behaviour of the compound with respect to determination is *mutual*.

The foundationalist might hold that there is a different reading of 'the behaviour of the parts'. It is not the behaviour of the particles considered on their own but, rather, the particles' actual behaviour in the field that is generated by the other particle. (The other particle itself is not part of the subsystem.) Thus, the behaviour of the first subsystem consists of the first particle's behaviour in an external gravitational field generated by the second particle. The second subsystem is described analogously. The two subsystems behave according to the Hamilton equations with the following Hamilton functions:

$$H_{1*} = p_1^2/2m_1 - (Gm_1 m_2/r) \big|_1$$
$$H_{2*} = p_2^2/2 m_2 - (Gm_1 m_2/r) \big|_2$$

'$|_i$' indicates that the function $(Gm_1 m_2/r)$ is restricted to the phase space of particle i. Let me stress that I am not committed to the claim that this can be consistently done in general. The foundationalist who takes this option is confronted with a dilemma here: Either the particle's actual behaviour (i.e., the particle's behaviour in the external field) cannot be individuated as indicated above, in which case it is not clear in what sense part-whole explanations commit us to ontological priority, because it remains unclear what the parts' behaviour is; or there is some way of individuating the parts' behaviour in this sense, but then it wouldn't help the foundationalist's argument. What we would end up with is a Hamiltonian that has the same form as in the non-interaction case:

$$H_{1+2} = H_{1*} + H_{2*}$$

So, by the same kind of argument as in the non-interaction case, the dependence relation would turn out to be mutual.

To sum up: Our starting point was the question of whether the reductive practice of part-whole explaining the behaviour of compound systems commits us to Foundationalism. As I argued at the outset of the chapter, Foundationalism requires – among other things – that an ontological priority relation obtains: Different kinds of facts/entities/events can be ordered such that some are ontologically more fundamental than others. Furthermore, I argued that for a relation to be an ontological priority relation it has to be non-modal and asymmetric. We have granted that part-whole explanations need a dependence or backing relation to work. But a commitment to a dependence relation does not commit us to an ontological priority of the behaviour of subsystems vis-à-vis the behaviour of the compound system (the obtaining of an ontological priority relation is – as I have argued – a necessary condition for the truth of Foundationalism). Physical part-whole explanations work in virtue of underlying dependence relations between the behaviour of the compound and that of the parts. These dependence relations are described in terms of laws of composition, which are laws of mutual dependence and thus not asymmetric. In other words, the metaphysical commitments of Chapters 1 and 2 suffice to account for the backing relation that has to be presupposed in part-whole explanations. *Inheritance* is false not only as a general claim but also in the special case of part-whole explanations.

One puzzle remains. What about the asymmetry of physical part-whole explanations? We do explain from parts to wholes, not vice versa. If the underlying dependence relation fails to be asymmetric, where does the asymmetry come from?

We have seen that an explanation may be asymmetric, e.g., when we explain a change in current in terms of a change in voltage but not vice versa, even though the underlying dependence relation (as described by Ohm's law) might well be symmetric with respect to determination. The asymmetry of the explanation in such a case is not due to the underlying dependence relation but, rather, due to the fact that we manipulate the one rather than the other (or that we are interested in what happens with current if we manipulate voltage).

In the case of part-whole explanations, we have seen that the non-asymmetric laws of composition work as the backing relation. And there might very well be a story that appeals to asymmetries in the pragmatic context that accounts for the asymmetry of part-whole explanations.

Consider causal explanations first: We are interested in causes to a large extent because they allow us to manipulate systems. Causes give us handles to bring about the changes we are interested in. Causal explanations trace the dependence relations that are required for such manipulations. Similarly, we can give a pragmatic account for our interest in constitutional relations: They give us additional handles to manipulate the behaviour of systems.[6] Having a grip on the behaviour of the parts of systems gives us – at least potentially – fine-grained handles to bring about the changes we wish to make. Part-whole explanations trace the dependence relations that are required for such manipulations. The asymmetry of explanation might mirror our interest in manipulating the compound by manipulating parts, which usually allows very fine-grained manipulations, while we are not very often interested in manipulating a particular part by manipulating the compound and holding fixed all the other parts.

What I have *not* given in this section is an argument to the effect that an ontological priority relation *cannot* obtain between parts and wholes. That was not my aim. Rather, what I have argued is that the part-whole explanations do not *commit* us to assuming that an ontological priority relation obtains between parts and wholes.

Let me add another point: We have seen that we can account for the dependence/backing relation in part-whole explanations in terms of laws of composition. This implies that the dependence relation in question holds with nomological necessity. In the light of our analysis of nomological necessity in Chapter 1, we have thus accounted for the backing relation in part-whole explanations in terms of invariance, which is a modal notion.

So, in the end, there are two reasons to shun the argument from part-whole explanations to ontological priority and thus to Metaphysical Foundationalism: Part-whole explanations do not commit us to holding that the dependence/backing relation is asymmetric, nor do they commit us to holding that the backing relation is non-modal.

6.4 Generality, Supervenience and Ontological Priority

Here is another option for the foundationalist. As we have seen in Chapter 5, it is plausible that physics attempts to explain the behaviour of *every* compound system in terms of the behaviour of parts. Given the

---

[6] I try to give a *general* account of why we might be interested in constitutional relations. This is meant to be compatible with the fact that often the manipulation of the parts is technically not feasible or sometimes – as in thermodynamics – we are only interested in manipulating coarse-grained variables.

rationale we have given for part-whole explanation, it is reasonable to assume that physics is aiming at *general* theories that pertain to every single system in the universe. If a theory is general, we should (at least in principle) be able to describe, explain and predict the behaviour of every single system (on whatever scale) in terms of this theory.

Given this notion of generality, what could an argument for the obtaining of an ontological priority relation look like? For an argument from generality to Foundationalism to work, there needs to be a premise that links descriptive generality and ontological priority. Suppose that we can describe, explain and predict everything in terms of physics but that we cannot explain or describe everything in more macroscopic terms. Every change in the behaviour of a system accounted for in terms of the less general theory will also be accounted for in terms of the general theory. What does that tell us about ontological priority? Plausibly, generality in the sense outlined commits us to supervenience:

If one theory A is more general and another theory B is less general (and the domain of B is subsumed by the domain of A), then it is implied that the behaviour of B-systems supervenes on the behaviour of A-systems.

However, supervenience does not give us what we are looking for. I can be very brief because the relevant facts are fairly uncontroversial. First, supervenience, even though not a symmetrical relation, is not an asymmetrical relation either, while ontological priority, as we have seen, is an asymmetrical relation. Furthermore, even though there are many different versions of supervenience discussed in the literature, it is generally taken to be a purely modal notion (see McLaughlin and Bennett 2018).

Thus, the generality of theories might commit us to the claim that the facts, etc. described by the less general theories *supervene* on the facts, etc. described by the more general theory. This commitment, however, is not a commitment to the assumption that an ontological priority relation obtains between these facts, etc. and is thus not a commitment to Foundationalism.

Against this line of argument, it might be objected that while we are not *committed* to Foundationalism, Foundationalism might still provide the best (or simplest) explanation of why the behaviour of compound systems supervenes on that of the parts. If A grounds B, then an explanation involving A and B is as parsimonious as one only involving A (only the fundamental things count as ontological commitments). Thus, the Foundationalist view is simpler than a non-foundationalist view. And, we should infer the simpler explanation. I agree that one might try to establish Foundationalism via inference to the best explanation. However, I will argue in Chapter 7 that Foundationalism does not provide the best explanation.

6.5   Objections and Replies

For the foundationalist, there are various possible ways to react to the argument just presented. First, one might object to it by pointing out that there might be genuinely metaphysical relations that obtain between the behaviour of parts and the behaviour of wholes but are not dealt with in physics. Answer: While there might be such relations, they are not my concern in this chapter. My aim is to develop a *minimal* metaphysics for scientific practice and thus to figure out whether Foundationalism is needed to account for how we use the part-whole relation in classical mechanics and quantum mechanics.

Second, one might argue that the equations that I relied on do not capture all that classical and quantum mechanics have to say about the part-whole relation. An analogous position is sometimes attributed to Cartwright with respect to causation (Field 2003, 443). However, while there is no a priori argument against this possibility, there is no account that I know of that tells us what additional physical facts there might be concerning the part-whole relation (that is, over and above those captured in the equations of classical or quantum mechanics). In the absence of such a positive account, it is difficult to evaluate this objection.

Finally, and perhaps most importantly, one might doubt that the picture I have presented is what physicists had in mind when they claimed that the behaviour of the whole is determined by the behaviour of the parts in an asymmetrical way. After all, we are dealing with microscopic physics, not just with two or three particles. So, the objection is to point to further dependence relations between the behaviour of parts and the behaviour of wholes that I have not addressed but that we are committed to in part-whole explanations. The objections dealt with in the following sections, in particular those in Sections 6.5.2 and 6.5.3, will consider the possibility of further candidates for the *in virtue* relation.

6.5.1   *Flagpole*

In the literature on explanation, there is the well-known case of the height of a flagpole and the length of its shadow. According to the laws of geometrical optics, the length of the shadow is determined by the height of the flagpole, holding fixed certain circumstances like the position of the sun. At the same time, these circumstances plus the length of the shadow determine the height of the flagpole. So, we have a case of mutual determination. With respect to this determination relation, the principle of asymmetry does not hold. Nevertheless, we do believe that the fact that the shadow has a certain length obtains partially in virtue of the fact that the flagpole has a certain length but not vice versa. By analogy, even though the determination relation between parts and wholes might fail to obey the principle of asymmetry, it might still be true that the behaviour of the compound obtains *in virtue* of the behaviour of the parts.

The reply is that the two cases are disanalogous in a relevant way. In the case of the flagpole, we can give an account of how the asymmetry arises, whereas we cannot do the same for the relation of parts and wholes.

Here is one way of explaining the origin of the asymmetry in the case of the flagpole. Geometrical optics is a simplified model of the situation at hand. A more detailed description would mention the propagation of light waves. In the more complete picture, it is possible to explain in what sense the length of the shadow is the dependent variable. Gerhard Schurz suggested that what's essential in this context is the fact that a change in the dependent variable is brought about *later*:

> The crucial idea [...] is that the distinction between those variables which are directly influenced by an allowed intervention, in contrast to those which are only indirectly influenced by it, is possible by considering the delays of time in the process of disturbing the system's equilibrium state. (Schurz 2001, 61)

And with respect to our example:

> Hence in every intervention allowed by C [circumstances like the position of the sun] which disturbs the equilibrium state of the system's variables, the length variation of the shadow will take place slightly after the variation of the pole's length – because of the finite velocity of light. (Schurz 2001, 61)

I will not discuss whether this suggestion does indeed give a complete account of the asymmetry in this example. The essential point is that this strategy to break the symmetry cannot be applied in the case of parts and wholes. What is relevant for Schurz's strategy is that we supplement the original description of the relation of the length of the shadow and the height of the flagpole by *additional physical facts* such as the propagation of the light wave. The simultaneous and mutual determination of the height of the flagpole and the length of its shadow is only apparent. It is merely a feature of a simplified and incomplete description of the situation. Breaking the symmetry relies on a better and more detailed description.

However, the case of parts and wholes is different in this respect: There are no additional physical facts. For all we know, the description of the part-whole relation given in Section 6.3 is the most complete one we have.

6.5.2 One-to-Many Relation

Even though our account of the part-whole relation as described in classical and quantum mechanics may be complete, the account may still give room for the obtaining of asymmetries that have been overlooked so far.

Frank Jackson argues that the asymmetry characteristic for physicalism is due to an asymmetry of determination: For the physicalist, the asymmetry between physical and psychological (or semantic or economic or biological ...) lies in the fact that the physical fully determines the psychological (or semantic ...), whereas the psychological (or semantic ...) grossly underdetermines the physical (Jackson 1998, 15).

An analogous argument in the case of Foundationalism runs as follows: While the behaviour of the parts fully determines that of the compound, the behaviour of the compound grossly underdetermines that of the parts. In other words: The relation between the whole and the parts surely seems asymmetrical in that a certain behaviour of the whole (dynamic or state) corresponds to many different arrangements of the parts.

However, as I will argue, this one-to-many relation does not suffice to establish an asymmetry claim. Let me illustrate this with a simple example. Suppose we are dealing with a massive compound system consisting of three subsystems. We are only interested in mass. Leaving out relativistic effects, we know that the mass of the compound (m_4) adds up as follows:

$$m_1 + m_2 + m_3 = m_4 \tag{M}$$

Thus, (*M*) is our law of composition for our three masses. m_4 characterises the compound or macro-system, whereas m_1 to m_3 characterise the constituents or micro-systems. Let us assume that the compound system has a mass of 17 kg. This value is compatible with a plethora of values for m_1 to m_3. 1 kg/5 kg/11 kg, 6 kg/6 kg/5 kg, 7 kg/6 kg/4 kg – all these micro-states are compatible with a macro-state of 17 kg. We have a one-to-many relation between the compound and its constituents, which seems to support an asymmetry claim and therefore (perhaps) the obtaining of an *in virtue* relation, asymmetry being a necessary condition for the obtaining of an *in virtue* relation. However, the same kind of one-to-many relation occurs if we fix a value for one of the constituents, say m_1. If m_1 is fixed at 5 kg, that is compatible with a plethora of values for m_2 to m_4: 5 kg/5 kg/15 kg, 6 kg/6 kg/17 kg, 3 kg/7 kg/15 kg – all of them will do. The fact that the compound has a certain mass value is compatible with lots of value distributions for the subsystems, but that does not single it out as something special.

The laws of composition give rise to equations that allow the behaviour of the compound to be calculated on the basis of the behaviour of the constituents. (Calculation presupposes determination of the relevant magnitudes.) However, they equally allow the behaviour of a constituent to be calculated, given the relevant information about the compound and the other constituents. Whenever we have three values in (M) we can calculate the fourth value. In this respect, there is nothing special about m_4, the value for the macro-state; with respect to determination, all the values are on a par. In this sense, the laws of composition (in quantum mechanics as well as in classical mechanics) are impartial with respect to the micro and the macro. It is true that the behaviour of the parts fully determines that of the compound and the behaviour of the compound grossly underdetermines that of the parts. However, it is also true that the behaviour of the first and second parts together with that of the compound fully determines the behaviour of the third part, while the third part on its own grossly underdetermines the rest.

6.5.3   Coarse Concepts

When it comes to the thermodynamics of, say, ideal gases, we do not only encounter the one-to-many relation as discussed in the previous section. There seems to be a further candidate for an asymmetrical relation.

The macro-description in terms of pressure (p), volume (V) and temperature (T) plus the exact specification of N – 1 particles doesn't determine the state of the 'last' particle (the Nth particle). There are various possible states that are compatible with the given constraints. On the other hand, the specification of all particles does determine the values for p, V and T. Is that an asymmetrical relation of the relevant kind?

For example, if temperature is mean kinetic energy, the velocities and positions of N – 1 particles and the temperature of the gas don't determine the velocities and position of the Nth particle. There is a whole set of velocities of the Nth particle compatible with a certain temperature of the gas plus the velocities and positions of the N – 1 particles.

Rejoinder: For a start, I will leave out the thermodynamic description of the ideal gas and focus on the mechanical description. Let's assume we have a complete description of the compound system (the gas). The state of the compound can be represented as a point in 6 N-dim phase space. Given the state of the compound as well as the states of N – 2 parts, the state of the second to last particle is not yet completely determined, because it can get into either the N – 1 slot or the N slot. However, given the state of the compound and the states of N – 1 particles, the state of the Nth particle is determined. Of course, the particles' states also determine the state of the compound. In this sense, we have mutual determination of parts and wholes at the level of a purely mechanical characterisation.

When we describe the ideal gas in terms of thermodynamic properties such as temperature and pressure, we use a coarser description of the compound system. It is coarse in the sense that a lot of micro-states are compatible with given values for p, V and T. Because we use this coarse terminology, i.e., p, V, T, for the compound system, the states of N – 1 particles plus the state of the compound fail to determine the state of the Nth particle. Strictly speaking, this is a case in which the variables representing the behaviour of the compound system are determined by the variables representing the behaviour of the parts, whereas it does not hold that the variables representing the behaviour of N – 1 parts plus the variable(s) representing the behaviour of the compound determine the variable for the Nth particle's behaviour.

However, I think we have good reasons not to take this asymmetry at face value, i.e., not to read it, as in fact, telling us something about the underlying ontology. The reason is that the asymmetry is generated by our choice of coarse-grained variables for the compound system. The asymmetry disappears if we choose the more precise mechanical description. Furthermore, asymmetries that are due to coarse-grained variables can be generated at will. This can be illustrated by the following example: Let's define an object as *heavy* if it weighs, say 150, 151, . . . or 200 kg. If the object has N parts, then the masses of the N parts determine whether or not the object is heavy. But the object being heavy plus the masses of N – 1 parts do not determine the mass of the Nth part. The parts determine the whole, but the whole plus N – 1 parts do not determine the remaining part.

However, the same kind of coarse concept can be defined for one of the parts. Take part no. 7. Part no. 7 is *quite heavy* if it weighs 50 or 51 or 52 or 53 kg. If the compound that no. 7 is a part of has N parts, then the mass of the compound plus all the masses of the other parts determine whether or not no. 7 is quite heavy. However, the mass of the compound is not determined by no. 7 being quite heavy plus the masses of the other parts (because of the coarseness of 'quite heavy').

What this shows is that we can generate asymmetries at will wherever we introduce coarse-grained variables. Therefore, we should not read these asymmetries as real. They are entirely due to the choice of coarse rather than precise variables and do not seem to have any implication with respect to the question of what kind of ontological relations obtain between parts and wholes.

6.6   Constraint and Ontological Priority

Up to this point, I have examined whether our reductive practice concerning part-whole explanations commits us to the obtaining of an ontological priority relation in nature. I have argued that this is not the case. In this section, I will discuss whether, if we additionally look at theory reduction, the picture changes.

At the outset of Chapter 5, I quoted Hoefer and Smeenk's definition of fundamentalism:

> Fundamentalism: there is a partial ordering of physical theories with respect to 'fundamentality'. The ontology and laws of more fundamental theories constrain those of less fundamental theories; more specifically: (1) the entities of the less fundamental theory T_i must be in an appropriate sense 'composed out of' the entities of a more fundamental theory T_f, and they behave in accord with the T_f-laws; (2) T_f constrains T_i, in the sense that the novel features of T_i with respect to T_f either in terms of entities or laws, play no role in its empirical or explanatory success, and the novel features of T_i can be accounted for as approximations or errors from the vantage point of T_f. (Hoefer and Smeenk 2016, 117)

If fundamentalism in this sense holds, compound macroscopic (less fundamental) systems are composed of parts, which can be described in terms of the more fundamental theory, and this then accounts for the behaviour of the compounds.

I take it that Hoefer and Smeenk's first condition (HS1) requires that (i) a part-whole relation obtains between the entities on the more fundamental and the less fundamental level and (ii) the behaviour of the less fundamental system must be part-whole explained in terms of that of the parts, i.e., in terms of the more fundamental theory. We have already dealt with the ontological commitments of part-whole explanations in the previous sections. I will now focus on their second condition. According to Hoefer and Smeenk's second condition (HS2), the more fundamental theory T_f is supposed to *constrain* the less fundamental theory T_i. In their explication of condition (HS2) and in particular, of the claim that 'the novel features of T_i can be accounted for', Hoefer and Smeenk write:

> The empirical and explanatory success of T_i must be grounded in the fact that it captures important facts about the structures identified by T_f. Or, in other words, T_i's success should be recoverable, perhaps as a limiting case within a restricted domain, in terms of T_f's ontology and laws. (Hoefer and Smeenk 2016, 117)

Thus, what condition (HS2) requires is that T_f constrains T_i in the sense that the success of T_i needs to be accounted for from the perspective of T_f.[7]

For our purposes, the essential question is whether the satisfaction of the constraint condition implies the obtaining of an ontological priority relation between the behaviour of systems as described in terms of T_f and the behaviour of systems as described in terms of T_i.

What Hoefer and Smeenk have in mind by *constraint* and *recoverability* is limit-case reduction as discussed in Chapter 5. Limit-case reduction aims to explain the past success as well as the continued application of a superseded theory from the perspective of a successor theory. It is an *epistemic* requirement that the successes of the old theory should be recoverable from the perspective of the new theory.

It is an interesting question whether the initial theory constrains the more fundamental or vice versa. It seems that in the process of the construction or development of fundamental theories, the less fundamental theory provides a constraint for the candidates of the more fundamental theory. By contrast, once the new fundamental theory is established and accepted as true, it is conversely the fundamental theory that puts a constraint on the less fundamental theory (it is this latter situation that Hoefer and Smeenk have in mind).

In one of the examples we discussed in Chapter 5, the special theory of relativity (STR) puts a *constraint* on classical mechanics, in that, once STR has been accepted as true, classical mechanics is only acceptable if its successes can be accounted for by the fundamental theory. This is achieved by showing that the solutions to the dynamic equations of classical mechanics approach the solutions of STR in the limit of small

---

[7] Ladyman and Ross (2013, 138) argue that fundamental theories impose 'consistency restrictions on all hypotheses in all sciences'. On a strict, logical reading of 'consistency', because less fundamental theories typically make predictions differing from those of the fundamental theories (even though these predictions concern the same domain), this condition will hardly ever be satisfied. On a more charitable reading, what needs to be met is Hoefer and Smeenk's recoverability requirement.

velocities. We then have a justification for the continued use of classical mechanics.

What does that tell us about ontological priority? Not very much, it seems. To see that 'constraint' is an issue that is tangential to questions of ontological priority, it suffices to ask what the relata of the ontological priority relation might be in this case.

First, there is an initial difficulty that can be circumvented. When metaphysicians talk about Foundationalism, they talk about a relation that obtains between facts, entities or maybe events. By contrast, physicists and philosophers of science such as Hoefer and Smeenk are usually interested in whether disciplines, fields or theories are (more or less) fundamental. So, the relata of the respective claims are different, and it seems that the attempt to connect these discourses is doomed already at this point. However, it is not too difficult to establish a connection.

Suppose some theory is the fundamental theory (or more fundamental than others) in the sense of Hoefer and Smeenk. What we are interested in is the following question: Is it implied that the facts, entities, etc. described by the fundamental theory are metaphysically fundamental vis-à-vis facts, entities, etc. described by non-fundamental theories (or by less fundamental theories)?

However, even with this translation at hand, we will not be able to establish that the constraint condition commits us to ontological priority. The problem is that we are here dealing with at most one true theory. Let us assume that STR is true. Classical mechanics and STR make different predictions, e.g., for the kinetic energy of a system at $0.5\times$ the velocity of light. Classical mechanics is, strictly speaking, false. That is, of course, what gave rise to the recoverability constraint in the first place: We want to understand how a false theory was successful and why we should be allowed to continue to use it. The problem in our context, however, is that if classical mechanics is false, then there are no 'classical mechanical' facts, entities, etc. Rather, what we should say is that both STR and classical mechanics are descriptions of the same facts, entities, etc. STR makes better predictions and is therefore more likely to be true.

The constraint condition is an entirely *epistemic* commitment: If we have different characterisations of the behaviour of a system in terms of different theories (or of the behaviour of the whole, on the one hand, and the behaviour of its parts and their interactions, on the other), these descriptions have to fit together – at least within certain domains. This is what needs to be shown to satisfy the constraint condition.

In other words, if fundamentalism in the sense of Hoefer and Smeenk is true, the satisfaction of the constraint condition can be explained as an epistemic commitment. It does not commit us to postulate any metaphysical relations. Thus, Hoefer and Smeenk's constraint does not commit them or us to the claim that an ontological priority relation obtains between more and less fundamental facts, entities or events.

6.7   Conclusion

Foundationalism is not implied by what classical mechanics and quantum mechanics have to say about the part-whole relation.[8] Not even those cases in classical and quantum mechanics that are most favourable to the foundationalist – namely, cases of part-whole explanation of the dynamics of compound systems – commit us to the claim that the behaviour of the compound is ontologically posterior to the behaviour of the parts (and some further facts about how the parts interact and how they are related). The reductive practices we have discussed in Chapter 5 do not commit us to Foundationalism.

---

[8] McKenzie comes to a similar conclusion: 'it seems that the layered structure that (those who identify as) metaphysicians are primarily concerned with is a structure that is oriented in some sense "orthogonally" from that relating the ontologies of the different sciences' (McKenzie 2019, 56). See also McKenzie (2011).

CHAPTER 7

Reduction and Ontological Monism

In Chapter 5 I looked at some reductive explanatory practices and argued that we can explain why we are interested in these reductive practices by appealing to the regulative ideal of achieving an understanding of how different descriptions of either one system (part-whole explanations) or classes of systems (theory reduction) are related. Still, explaining the success of these practices might commit us to either Foundationalism or Physical Eliminativism. In Chapter 6 I argued that Foundationalism is not implied by our scientific practice. In particular, there was no need to assume ontological priority relations to obtain. In this chapter I will examine two different brands of ontological monism – views that do without the assumption that there are different layers of facts/entities/events. Physical Eliminativism and Ontologically Neutral Monism agree that ontologically speaking there is no hierarchy of facts/entities/events. I will argue that from the perspective of a minimal metaphysics of scientific practice, in comparing these two views, Ontologically Neutral Monism should be preferred to Physical Eliminativism because the former can explain all the features of scientific practice that the latter explains while making fewer metaphysical assumptions.

7.1 Physical Eliminativism

In Chapter 5 we characterised Physical Eliminativism as the view that the only facts there are, are fundamental physical facts; macroscopic everyday facts are only apparent facts. All but the physical facts are eliminated.

Physical Eliminativism differs from Foundationalism because where Foundationalism sees two kinds or layers of facts/entities/events that are related by a relation of ontological priority, for instance, microscopic or physical facts, on the one hand, and macroscopic facts, on the other, Physical Eliminativism sees only one kind or layer of fact/entity/event, namely, physical facts/entities/events. Furthermore, Physical Eliminativism

186

has to be distinguished from what I will call Ontologically Neutral Monism. Both agree in denying what Foundationalism presupposes, namely, that there is a relation between different kinds or layers of facts, entities or maybe events. Both agree that there is only one kind or layer of facts/entities/events. While Physical Eliminativism holds that it is the *physical* facts/entities/events that are the only facts/entities/events, Ontologically Neutral Monism is not committed to this metaphysical privileging of the physical (or the micro). Ontologically Neutral Monism is merely committed to the claim that there is one kind or layer of fact/entity/event and allows there to be various theoretical accounts of such facts/entities/events, i.e., various theoretical accounts that pick out the same facts/entities/events. I will now examine two arguments for Physical Eliminativism that might initially be thought to be implied by scientific practice and argue that they do not show what they purport to show. I will turn to Ontologically Neutral Monism in Section 7.2.[1]

Before I start to examine causal overdetermination arguments it is helpful to distinguish various levels of descriptions of systems. In an ideal gas, for example, we have the molecules, their velocities, etc. and their interactions – let's call this the description of the behaviour of the parts. Second, we have the system as a whole constituted out of these molecules and we may attribute a certain behaviour to this whole, e.g., a certain mean kinetic energy – call this the 'micro-based' description of the system. Furthermore, there is an additional description or characterisation in terms of thermodynamics; we might want to attribute a certain temperature to the gas – call this a macro-description of the gas (see Hüttemann 2004, 24–9 for an extended discussion of these distinctions). So, there are two different relations we should keep separate in our analysis:

1. The relation between the properties of the parts and the interactions of the parts, on the one hand, and, on the other hand, the micro-based description of the compound (e.g., the kinetic energy of the compound). The relevant question in this context is: How is the micro-based behaviour of the compound related to the behaviour of the parts?

[1] I hope that by calling this position 'Ontologically Neutral Monism' I do not cause too much confusion. There is no intention to revitalise Mach's Neutral Monism or other views with similar labels. The reason for calling the advocated position 'Ontologically Neutral Monism' is that – in contrast to Physical Eliminativism – a positive claim about the nature of the facts/entities/events that constitute the world is not built into the position.

2. The relation between a micro-based property of the compound (e.g., kinetic energy) and macroscopic properties (e.g., temperature). The relevant question in this context is: How are the two different characterisations of the compound (e.g., in terms of statistical mechanics and in terms of thermodynamics) related?

Physical Eliminativism may concern the first issue. According to this version of Physical Eliminativism, the only facts/entities/behaviours there are, are those concerning the (physical) parts of compound systems. Both the micro-based and the macro-based description of the compound fail to pick out genuine facts/entities/behaviours. It is, to return to our example, an eliminativism with respect to temperature as well as mean kinetic energy. A second, weaker version of Physical Eliminativism merely claims that the macro-properties have to be eliminated and is silent about the relation of the micro-based properties/behaviours to the properties/behaviours of the parts. The first version of Physical Eliminativism implies the second and is thus stronger.

This distinction between a weak and a strong version of Physical Eliminativism can be drawn provided we have (i) an account of how the parts and their interactions give rise to the behaviour of a compound that can be characterised within the same theory/vocabulary and (ii) a further account of the behaviour of the compound in terms of a different theory/vocabulary. A further illustration of these distinctions is pain, which we can describe both in mentalistic vocabulary and in terms of neurobiological processes, where the latter in turn can be explicable in terms of parts and interactions that constitute the process in question.

7.1.1   Causal Overdetermination

Causal overdetermination arguments can be used to argue for a variety of philosophical positions. I will start with the discussion of an argument that aims to establish that only the (ultimate, physical) parts exist but not compound beings. I will thus be concerned with the strong version of Physical Eliminativism.

The argument allegedly relies on our practice of causal explanation: Given the way we causally explain in the sciences, it is argued, we have good reasons to suppose that it is only parts rather than compounds that are causally relevant for effects. This in turn may serve as the basis for the claim that the behaviour of the compound is ontologically derivative (see, e.g., Trogdon 2018) or for the claim that we have no reasons to assume that compounds exist. This latter claim, i.e., strong Physical Eliminativism, has been defended by Trenton Merricks.

The causal overdetermination argument according to Merricks runs roughly as follows: Consider a compound system such as a billiard ball *A* bouncing into another billiard ball *B*, which then starts to move. Physics tells us that the parts of the billiard ball *A* cause billiard ball *B* to move. However, if *the parts* of *A in concert* cause the effect, there is nothing for *the compound* system to do. We should conclude that what the compound system appears to cause is epiphenomenal and caused only in virtue of what the parts cause. Everything that happens, happens in virtue of what the fundamental physical parts cause. (If one furthermore accepts the claim that existence goes with genuine causal powers, one might argue for what has been called 'nihilism' (van Inwagen 1990) or 'eliminativism' (Merricks 2001) about macro-physical objects: there are only suitably arranged fundamental physical parts and that's it.)[2]

I will now have a closer look at Merricks' overdetermination argument:

(1) Object O – if O exists – is causally irrelevant to whether its parts $P_1 \ldots P_n$, acting in concert, cause effect E.
(2) $P_1 \ldots P_n$ cause E.
(3) E is not overdetermined.

Therefore:

(4) If O exists, O does not cause E. (Merricks 2001, 79–80)

Let us consider the premises of the arguments. Premise (2) seems unproblematic. When one billiard ball bounces into another and causes it to move, we are probably willing to assert that the parts of the billiard ball in concert cause the second ball to move. That's at least what I will assume. According to premise (1), the compound O is causally irrelevant to whether its parts $P_1 \ldots P_n$, acting in concert, cause effect E. This premise has to be analysed carefully. Being causally irrelevant to whether its parts $P_1 \ldots P_n$, acting in concert, cause effect E, according to Merricks' use of 'causally irrelevant', implies all of the following and nothing more than the following: (a) O is not one of the parts that cause E, nor is it (b) a partial cause of E alongside the parts; it is (c) not an intermediate cause, i.e., it is not the case that the parts cause O to cause E, nor is (d) O a cause of the parts, which then in turn cause E (Merricks 2001, 58). This is the sense in which O is causally

---

[2] Van Inwagen (1990) and Merricks (2001) both believe that organisms exist. For the purposes of this chapter I will not take this positive existence claim into account.

irrelevant to whether its parts $P_1 \ldots P_n$, acting in concert, cause effect E. Premise (1) thus does not claim that O is causally irrelevant *tout court* (that would be a *petitio*), nor is it claimed that it is irrelevant (without the qualification 'causally') to whether its parts $P_1 \ldots P_n$, acting in concert, cause effect E. Read this way, premise (1) is not objectionable, either. The problem is with premise (3). What it effectively says is that if the parts of O cause E then it cannot possibly be the case that O causes E, too. What is it about causation that could warrant premise (3) in the case of the parts and the compound?

Here is one view that might make sense of premise (3): According to the sixteenth-century Jesuit Francisco Suárez, 'a cause is a principle inflowing by itself being to something else' [my translation] (Suárez, 1866, 384). Given Suárez's conception of causation as the transfer of being, it might seem plausible that there is a competition between the behaviour of the parts and that of the compound with respect to giving some sort of complete being to the second billiard ball's motion. Presumably, it can only be the case that *either* the parts are sufficient for giving complete being to the effect *or* the compound is sufficient for giving complete being to the effect, but it cannot, on this conception, be the case that *both* the parts *and* the compound are sufficient for giving complete being to the effect. If both were sufficient causes, we would have too much being, twice the effect or two effects.

But how about contemporary accounts of causation? The view that causation needs to be spelled out in terms of the transfer of being has long been dismissed. So, the question is whether contemporary accounts exclude the option that both the parts and the compound are sufficient causes of the effect in question. Suppose causation is spelled out in terms of a regularity view. Given premise (2), there is a regularity such that whenever the parts act in concert in a certain way, the effect occurs (nothing depends on the particular version of the theory). Is there anything in regularity theories that would disallow a further regularity between the compound object O's behaviour and the effect? As far as I can see, there isn't: the basic idea of a regularity account does not exclude the possibility that both the parts' behaviour and the compound's behaviour are sufficient causes of the effect in question.[3] On this account there is no causal competition between the behaviour of the parts and that of the compound,

---

[3] One might, of course, simply add by *fiat* a clause that disallows cases of causal overdetermination. Mackie in an early presentation of his INUS-theory of causation had such a clause (1965, 247). He later dropped it (Mackie 1980).

and thus, a prohibition of causal overdetermination in the case of the behaviour of parts and wholes cannot be made plausible.

Similarly, for counterfactual accounts of causation: Suppose there is a counterfactual dependence between the parts acting in concert and the effect E. This dependence does not per se exclude that there is furthermore a counterfactual dependence between the compound's behaviour and E. (A lot, of course, depends on the exact semantics for counterfactuals, and I am not claiming that, if there is a counterfactual dependence between the parts' behaviour and E, then it is guaranteed that there is a counterfactual dependence between the compound's behaviour and E (see, for instance, Block 2003 and Bennett 2008 for discussion).

Process theories of causation, which spell out causal processes and interaction in terms of conserved quantities, have often been taken to be more fitting for overdetermination arguments (see Kim 2002 and Loewer 2002). However, in this case they are of no help either. Physics allows us to say that the parts in concert transfer a certain amount of energy and momentum to the second billiard ball. At the same time, we may say that the compound transfers a certain amount of energy and momentum to the second billiard ball. There is no competition here either.

Similarly, my own account of causation (see Chapters 3 and 4) in terms of quasi-inertial processes (in this case, the second billiard ball *B* being at rest) allows us to characterise both the movement of the compound and the movement of the parts in concert as the cause of billiard ball B to start moving.

The essential point is that contemporary accounts of causation – in contrast to a traditional account of causation such as Suárez' – do not imply that there is a competition between causation due to the behaviour of the compound and due to the behaviour of its parts (given that the behaviour of compound systems can be explained in terms of that of the parts).[4] What is happening can be characterised at the level of the parts or at the level of the compounds. Both characterisations might be true at the

---

[4] Kim, whose overdetermination arguments were hugely influential, moved from claiming 'there can be no more than a single *complete* and *independent* explanation of any one event and we may not accept two (or more) explanations of a single event, unless we know, or have reason to believe, that they are appropriately related – that is related in such a way that one of the explanations is either not complete in itself or dependent on the other' (Kim 1988 [2010], 159), where he stresses that the explanations have to be *independent* to compete with one another, to a version of causal exclusion arguments where the requirement of independence is dropped, e.g., 'Exclusion: No single event can have more than one sufficient cause occurring at any given time – unless it is a genuine case of causal overdetermination' (Kim 2005, 42). While the initial claim seems entirely plausible to me, it is hard to see why the exclusion principle without the independence clause should be convincing.

same time. The only constraint is that the descriptions have to yield the same empirical predictions – they have to fit together.

To return to the main argument of this section: None of the currently held theories of causation warrants premise (3) as applied to compound systems and their parts. And, in fact, Merricks does not appeal to any of these theories but, rather, to intuitions we have about overdetermination, which might very well be informed by older ways of conceptualising causation. Additionally, he appeals to

> the 'scientific attitude' and 'bottom-up metaphysics' according to which the final and complete causal stories will involve only the entities over which physics quantifies. (Merricks 2001, 60)

This motivation for his argument – despite appearances – cannot draw on scientific practice. As I have just argued, there is no reason to assume that physics quantifies only over the parts of billiard balls but not over billiard balls.

Furthermore, in the context of our search for an argument for Physical Eliminativism an appeal to 'bottom-up metaphysics' amounts to a *petitio principii*.[5] In fact, it is precisely the prejudice that a 'scientific attitude' requires a bottom-up metaphysics that I intend to repudiate in Chapters 6 and 7.

To sum up: The overdetermination argument for eliminativism about macro-physical objects fails to convince. There is no argument from causal overdetermination to strong Physical Eliminativism. We cannot use the causal overdetermination argument for establishing strong Physical Eliminativism without already assuming 'bottom-up metaphysics', i.e., without begging the question.

Exactly the same kind of causal overdetermination argument that Merricks discusses with respect to molecules (and their behaviour) and the compound that is constituted by the molecules (and its behaviour) can be advocated with respect to the issue of how the micro-based behaviour of the compound is related to the macro-behaviour of the very same compound. That is, of course, the well-known overdetermination argument as presented most notably by Kim. If successful, it would be an argument for the weaker version of Physical Eliminativism.

---

[5] Merricks briefly considers that the prohibition of causal overdetermination does not apply when the causes aren't wholly separate but offers only a very lame reply: He argues (correctly) that with the assumption that overdetermination is problematic only if the causes in question are wholly separate, Kim's overdetermination argument wouldn't work because supervenient properties aren't wholly separate from their subvenient basis. (Merricks 2001, 71).

However, Kim's argument can be rejected for exactly the same reasons as Merricks'. There is nothing in contemporary theories of causation (with one possible exception) that would preclude the ascription of causal relations to either the micro-based properties or the macro-properties. The exception in this case (as opposed to the one we discussed earlier) is a process theory that spells out the relevant processes in terms of conserved quantities or other features that, arguably, only make sense within a micro-based description but not in a description that relies on macro-properties. However, I have already argued in Chapter 3 why we should reject this version of a process theory. On all other accounts of causation, such as regularity theories, counterfactual theories and my own version of a disposition-based process theory, it does make sense to attribute causal relations as relating both to micro-based behaviour and to macro-behaviour.[6]

7.1.2 *Supervenience (Again)*

In Section 6.4 we argued that we may have good reasons to assume that physics is aiming at *general* theories that pertain to every single system in the universe, that is, at theories that allow us (at least in principle) to describe, explain and predict the behaviour of every single system (on whatever scale) in terms of this theory. Generality in turn commits us to supervenience:

> If one theory A is more general and another theory B is less general (and the domain of B is subsumed by the domain of A), then it is implied that the behaviour of B-systems supervenes on the behaviour of A-systems. (I will understand supervenience somewhat loosely as saying that there are no B-changes without A-changes.)

It was pointed out that while supervenience does not commit us to an ontological priority relation and thus to Foundationalism, it might still be the case that Foundationalism provides the best explanation .for why B-facts supervene on A-facts. It might be argued that if A-facts ground B-facts, i.e., if A-facts are ontologically prior to B-facts, then there can be no change in B-facts without a change in A-facts. So, Foundationalism

---

[6] There is an extended debate about whether the overdetermination argument can be run within the interventionist picture. The main problem concerns the question of whether, if causes depend on each other via, e.g., supervenience, they can be integrated into one causal model; see, e.g., Baumgartner 2010, Hoffmann-Kolss 2014 and Woodward 2015.

does provide an explanation of the supervenience that comes with facts described by more general and less general theories that typically play a role in reductive practices.

But is this the best explanation of why the behaviour of compound systems supervenes on that of the parts, their interactions, etc.? Physical Eliminativism might provide a better explanation – at least if we allow ourselves to add considerations of parsimony. The central idea is that having both a billiard ball and a multitude of molecules arranged ball-wise (which are related by a relation of ontological priority) does not do any work in explaining scientific practice that could not be done by the multitude of molecules arranged ball-wise alone.

Physical Eliminativism can explain the supervenience of the behaviour of compound systems on that of the parts, their interactions, etc. as follows. Physical Eliminativism assumes that it is only the physical entities/behaviours that exist; there are no other (macroscopic) entities/behaviours. So, according to Physical Eliminativism, if a more general theory allows to describe the behaviour of (ideally) every system in every detail (the A-theory) and a less general theory describes it in less detail (the B-theory), then both theories describe the same facts, namely, A-facts (which, because fundamental physics provides the most general theories, will be the fundamental physical facts). Thus, if the A-theory is more (or most) general and the A-facts are the only facts around, there can be no B-changes (that is, changes described by the B-theory) without A-changes. The reason is that the changes described by the B-theory, according to Physical Eliminativism, simply are A-changes.

So, Physical Eliminativism does provide an explanation of why the behaviour of compound systems supervenes on that of the parts, their interactions, etc. The explanation is better than that provided by Foundationalism because – while accounting for exactly the same explanandum – fewer metaphysical assumptions are necessary to provide the explanation. Two sets of entities/behaviours that stand in a relation of ontological priority do no more work than the Physical Eliminativist's set of (fundamental) physical facts, properties, etc. However, even though Physical Eliminativism provides a better (more parsimonious with respect to metaphysical assumptions) explanation of supervenience than Physical Foundationalism, it might not be the best explanation around. In the next section I will argue that there is a still better explanation. Our analysis of scientific practice thus does not commit us to Physical Eliminativism.

7.2 Ontologically Neutral Monism

As I have argued in the previous sections, scientific practice commits us neither to Physical Eliminativism nor to Foundationalism. However, there are positive commitments that we have examined in the previous sections. Let me briefly recapitulate these commitments.

First, in many cases of part-whole explanation it is presupposed that compounds exist and that they have parts. In these cases, a modal, asymmetric relation of generic existential dependence (not between compounds and their parts but) between compounds and kinds of parts has to be presupposed. A particular physical compound does not depend *for its existence* on the *existence* of a particular part. It does, however, depend for its existence on there being certain kinds of parts, and this is what was called 'generic existential dependence' (Section 6.2). Furthermore, this dependence is asymmetric. While the existence of certain sorts of entities (parts) is necessary for the existence of a particular physical compound, the converse is not true. However, this dependence relation does not suffice to establish ontological priority or Foundationalism, as we have seen in Chapter 6.

Second, we turned from the question of how the *existence* of a compound depends on the *existence* of kinds of its parts to the question of how the *behaviour* of an existing compound depends on the *behaviour* of its parts. In discussing the explanatory mechanism of part-whole explanations we have accepted explanatory realism and its commitment to so-called backing relations. Part-whole explanations presuppose that the behaviour of a particular whole (what is mentioned in the explanandum) depends on the behaviour and the interactions of its particular parts (i.e., on those features the explanans refers to). It is *laws of composition* that completely describe these dependence relations. As a consequence, it is not necessary to introduce further metaphysical inventory over and above what we introduced in Chapter 2 in the context of laws of nature in order to give an account of the backing relation in part-whole explanations.

So, the overall argument is that reductive explanations show that science requires a backing relation, and the backing relation is not required to be asymmetric; it need not be non-modal. We are thus not committed to Foundationalism. Nor are the arguments for Physical Eliminativism convincing, as we have seen in the previous section. However, while we are not *committed* to one of these views, we might still have good reasons to embrace one of them because it provides the best explanation of some aspect of scientific practice or some other feature. For instance, both

Foundationalism and Physical Eliminativism metaphysically account for why the macro-behaviour (e.g., temperature) of some systems supervenes on its micro-based behaviour – the first in terms of grounding and the second in terms of identity. Supervenience does not *commit* us to one of them, because more than one account is available. Still, one of these accounts may provide a better explanation than the others. Via inference to the best explanation we might then hold true one of these accounts.

In Section 7.1.2 I have already argued that Physical Eliminativism provides a better explanation than Foundationalism. There is, however, a further option that is metaphysically more parsimonious with respect to the metaphysical assumptions it is committed to than Physical Eliminativism, an option I will call 'Ontologically Neutral Monism'. In contrast to Foundationalism, Ontologically Neutral Monism agrees with Physical Eliminativism that when we are confronted with cases of supervenience, as, for example, in thermodynamics and statistical dynamics, we should not assume there to be two sets of properties, for instance, microscopic or physical facts, on the one hand, and macroscopic facts, on the other. However, in contrast to Physical Eliminativism, Ontologically Neutral Monism is not committed to metaphysically privileging the physical (or the micro). Ontologically Neutral Monism is merely committed to the claim that there is one kind of fact/entity/event and allows there to be various theoretical accounts of such facts/entities/events. On this basis the following explanation of the supervenience that comes with more and less general theories is available. According to Ontologically Neutral Monism, if there is a more general theory that allows us to describe the behaviour of (ideally) every system in every detail (the A-theory) and a less general theory that describes it in less detail (the B-theory) then both theories describe the same facts. Thus, if the A-theory is more (or most) general, there can be no B-changes (that is, changes described by the B-theory) without A-changes. The reason is that the changes described by the B-theory, according to Ontologically Neutral Monism, are the very same changes that the A-theory describes. The difference between Physical Eliminativism and Ontologically Neutral Monism is that while both hold that there is only one set or layer of facts/entities/events that are described by both theories, the former additionally postulates that the physical (or fundamental or microscopic) description is metaphysically privileged.

Thus, Ontologically Neutral Monism provides a more parsimonious account of our reductive practice than Physical Eliminativism because it does not insist on claiming that it is exactly one description of the behaviour of the system (namely, the (micro-)physical) that gives us the only true account of what is going on. There can be more than one true story of the behaviour of a system *provided these different accounts can be related appropriately, e.g., by reduction.*

In the remainder of this section I will illustrate what Ontological Neutral Monism amounts to by discussing a number of cases.

(i) *Billiard ball*: In the previous section we have discussed two causal accounts of a billiard ball that is bumping into, say, another billiard ball. This is a simple case that serves ideally to illustrate what Ontological Monism amounts to. We have two options. We can either attribute the momentum and energy in terms of which we can explain that the second billiard ball starts to move to the billiard ball considered as a compound (macro-description). Or – second option – we can attribute it to the parts of the billiard ball, which are suitably arranged and interact in certain ways (micro-based description). There is no need to postulate two different objects and to wonder about their relation. There is just one object (which on one account is picked out as 'the billiard ball' and on the other account as 'the parts of the billiard ball, which are suitably arranged and interact in certain ways') whose behaviour can be characterised in terms of the macro-description or in terms of the micro-description. The fact that we have a part-whole explanation shows that both descriptions are equivalent. Both accounts can be taken to be true.

(ii) *Eddington's table*: In a well-known passage the physicist Arthur Eddington ponders over what he is sitting at.

> I have settled down to the task of writing these lectures and have drawn up my chairs to my two tables. Two tables! Yes; there are duplicates of every object about me – two tables, two chairs, two pens. [...] One of them has been familiar to me from earliest years. It is a commonplace object of that environment which I call the world. How shall I describe it? It has extension; it is comparatively permanent; it is coloured; above all it is substantial. [...] Table No. 2 is my scientific table. It is a more recent acquaintance and I do not feel so familiar with it. [...] My scientific table is mostly emptiness. Sparsely scattered in that emptiness are numerous electric charges rushing about with great speed; but their combined bulk amounts to less than a billionth of the bulk of the table itself. (Eddington 1964, 5/6)

Eddington then goes on to argue that only one of these tables is real:

> I need not tell you that modern physics has by delicate test and remorseless logic assured me that my second scientific table is the only one which is really there – wherever 'there' may be. (Eddington 1964, 8)

According to Eddington (in the context of the passages quoted here)[7] fundamental physics gives us the one real story of what is going on in the world. It is modern physics itself, he suggests, which tells us that physics gives us the one true story and that Physical Eliminativism is true.

In the previous sections I have rejected this claim. Scientific practice does not commit us to Foundationalism, nor does it commit us to Physical Eliminativism. But what shall we say positively about the problem of the two tables?[8] Here is what Steven French has to say about the solidity of the familiar table:

> As already noted this holds in virtue of the relevant physics as expressed by the Exclusion Principle and, more fundamentally, the anti-symmetrization of the relevant aggregate wave-function. In this case one might then insist that the latter feature of quantum mechanics entirely explicates the solidity of everyday objects and in doing so eliminates the predicate from the scope of our fundamental ontology. (French 2014, 170)

French thus endorses Physical Eliminativism. But Ontologically Neutral Monism rather than Physical Eliminativism provides a better explanation of what is going on. We should not expect there to be two tables; that can be granted. But there is no 'remorseless logic' that forces us to conclude that fundamental physics gives us *the only* true story of the world. It is precisely because we have a reductive account of solidity in terms of quantum mechanics that we can conclude that we have two adequate and true accounts of a property of the chair. What is characterised as the solidity of the chair can at the same time be characterised in terms of quantum mechanics. We might privilege one or the other account in certain contexts but a minimal metaphysics of scientific practice does not commit us to an ontological privileging.

It might, of course, be the case that in developing reductive accounts we need to revise some of the assumptions we had. For instance, it might have been thought that the solidity of an everyday object requires that its volume is filled completely with material stuff all the way down. This assumption had to be revised, but that only means that we have learned something about solidity and does not make solidity go away. An account of the table in terms of macroscopic properties such as solidity and an account in terms of quantum mechanics can both be true. There is no reason to believe that alethic overdetermination is forbidden. An

---

[7] I make no attempt to do justice to Eddington's more elaborated views. See French (2014), 79ff. for discussion.
[8] There is a vast literature discussion on this issue; see French (2014), chapter 7.

ontological monist will argue that we have two true accounts of chairs and tables; reductive explanations make explicit why this possible. In different contexts there will be different pragmatic considerations that make us prefer the one or the other account.

Physicists are well aware of the fact that there can be different theoretical accounts of one and the same system, e.g., in terms of Hamiltonian and Lagrangian mechanics or in the context of so-called dual theories (see, e.g., Castellani and de Haro 2020).

(iii) *Phase transition*: In Section 5.6 we briefly discussed phase transitions and critical phenomena. Suppose we are dealing with some substance, say water, that undergoes a phase transition. We can distinguish three levels of description (as before): First, a description of the parts (molecules) on their own in terms of mechanics; second, a micro-based description of the substance as a whole in terms of statistical mechanics; and third, a description of the substance as a whole in terms of thermodynamics. According to Ontological Monism, two descriptions of the substance as a whole are descriptions of just one object. There are not two objects, nor is there a statistical-mechanical behaviour *plus* a thermodynamic behaviour. Both the thermodynamic account and the account in terms of statistical mechanics are characterisations of the behaviour of the substance in question.

As already indicated in Section 5.6, the behaviour of the compound if described in terms of thermodynamical quantities involves discontinuous changes in a derivative of a thermodynamic function, while the account in terms of statistical mechanics of finite systems involves no such discontinuities when it comes to the description of phase transitions. But no measurements with finite precision can establish that what we have measured needs to be described as truly discontinuous rather than as something smoother in the vicinity. In other words: As long as our measurements have only finite precision, we can account for the observed macro-behaviour both in terms of statistical mechanics and in terms of thermodynamics.

Both accounts give equally good accounts of the observed behaviour. Our reductive practices – in this case both part-whole explanations, which take us from the behaviour of individual molecules to the characterisation of the compound in terms of statistical mechanics, as well as inter-theory reduction, which allows us to see how and under what circumstances both theories converge on the same empirical predictions – let us understand why both theories give an empirically adequate account of what has been measured.

The ontological monist will hold that there are two theories that describe the behaviour of one and the same system. These theories may very well make different predictions, so that at most one of them can be true. This is a case of empirical underdetermination – at least at present.

But what do we make of the fact that statistical mechanics is the more *general* theory? Let us assume that statistical mechanics is the more general theory in the sense that the thermodynamic description is empirically adequate in a domain that is a strict subset of the statistical mechanics' domain of empirical adequacy.

Suppose T_1 gives us an empirically adequate account of systems x_1 and x_2, while T_2 gives us an empirically adequate account of x_2 only. What would be the argument for establishing that if the more general theory T_1 is true of x_2, therefore T_2 cannot be true of x_2 as well? If T_2 is false with respect to x_1 while T_1 is true with respect to x_1, this somewhat undermines our belief in T_2, I suppose. However, if we have a reductive account of why T_2 works for x_2 given certain conditions are met, there is no reason why the less general theory T_2 should not be true of x_2 too.

But what about the following thought? Theories typically postulate a mechanism in virtue of which they get things right or wrong. The more general theory postulates a widely successful mechanism while T_2 only got it right by accident – reduction tells us why T_2 *appears* to be right.

An interesting case is the second law of thermodynamics, which is part of a less general theory. This is clearly a case where attempts to reduce it to statistical mechanics have led to a reconceptualisation of what the law says (it is now considered to be a statistical law). Another interesting feature is that the pressure to come up with a good account of what is going on is on statistical mechanics rather than on thermodynamics. Even though the laws that go into statistical mechanics, and in virtue of which statistical mechanics is a general theory, are established beyond doubt (and provide a 'mechanism' to develop empirically adequate accounts of phenomena), these laws alone do not provide us with a reductive account of how the second law is connected to statistical mechanics. Much depends on assumptions about initial conditions, etc. These are non-general assumptions.

Thus, the mere fact that statistical mechanics is the more general theory, in the sense that the thermodynamic description is empirically adequate in a domain that is a strict subset of the statistical mechanics' domain of empirical adequacy, does not make the statistical mechanical account more truth-likely than the thermodynamic account. In fact, with respect to the case at hand, many have argued that the second law of thermodynamics is more likely to be true. Eddington, for example, noted:

> The law that entropy always increases holds, I think, the supreme position among the laws of Nature. If someone points out to you that your pet theory of the universe is in disagreement with Maxwell's equations – then so much the worse for Maxwell's equations. If it is found to be contradicted by observation – well, these experimentalists do bungle things sometimes. But if your theory is found to be against the second law of thermodynamics I can give you no hope; there is nothing for it but to collapse in deepest humiliation. (Eddington 1964, Chapter 4).[9]

(iv)   *Decoherence – classical properties*. In the case of chairs and tables, I have just argued, we learn something about their solidity when we turn to quantum mechanics. But sometimes when we turn to a new theory we learn quite a lot, and the acceptance of the new theory leads to a replacement of the old ontology.

Whether the acceptance of a new theory leads to learning more about the old objects and their properties (including some revisions) or, rather, to a replacement depends on how these objects are individuated, i.e., on what is taken to be *essential* for the objects in question. Learning that light is an electromagnetic wave or a bunch of photons does not commit us to give up our belief that light exists, because light was never individuated or defined in terms what it is composed of but (presumably), rather, in terms of laws, e.g., those of geometrical optics.

An interesting question is what we ought to say about classical objects and their determinate properties (such as being at a particular place or having this or that velocity) in the light of the decoherence approach that I briefly mentioned in Section 5.6. One possible conclusion is to argue that classical objects and their determinate properties are just illusions. It is essential to our conception of chairs and tables that they are localised and that their defining features are (maybe with the exception of their colour) non-relational, intrinsic properties. Quantum mechanics tells us that there are no such things. Furthermore, decoherence theory, our account of how

---

[9] Einstein similarly remarked:

> A theory is the more impressive the greater the simplicity of its premises, the more different kinds of things it relates, and the more extended its area of applicability. Therefore the deep impression that classical thermodynamics made upon me. It is the only physical theory of universal content which I am convinced will never be overthrown, within the framework of applicability of its basic concepts.(Albert Einstein 1979, 31, as quoted by Howard and Stachel, 2000, 1)

the classical and the quantum description are linked, explains *the appearance* of a classical property in terms of the interaction of the system in question with the environment (Joos et al. 2003; Schlosshauer 2004). What is explained is not why the compound has classical properties given the environment but, rather, why they *appear to be classical to an observer*. However, the conclusion that there are no chairs and tables need not be drawn provided they are individuated as functionally defined objects. Maybe we should not classify a chair as an object any more (depending on one's conception of object), but if a chair is simply that which we can sit on, the argument that quantum mechanics tells us these kinds of things don't exist, doesn't go through. Rather, thanks to quantum mechanics, we have simply learned a lot about their constitution and how and why they appear to us in the way they do.

To conclude, the positive picture that emerges is one that can be characterised in terms of 'ontological monism' and 'descriptive pluralism': It allows for a plurality of descriptions of a system (or of reality), none of which is ontologically privileged as the exclusively true account of reality, provided they are empirically adequate. However, these different descriptions – by virtue of being descriptions of one and the same system (or reality) – are constrained by an epistemic requirement: We need to be able to give reductive accounts of how and under what circumstances the different descriptions can be accounts of the behaviour of one and the same system.

CHAPTER 8

Concluding Remarks: Methods and Epistemic Sources in Metaphysics

In this final chapter I will briefly place the epistemic sources and the methods employed in the minimal metaphysics of scientific practice in the context of various other approaches to metaphysics and I will take up some of the methodological issues raised in the Introduction.

8.1 A Priori Metaphysics

According to one influential conception, metaphysics is a purely aprioristic endeavour. A prominent exponent of this conception was Kant. In his *Prolegomena* he writes about the sources of metaphysical knowledge:

> The very concept of metaphysics ensures that the sources of metaphysics can't be empirical. If something could be known through the senses, that would automatically show that it doesn't belong to metaphysics; that's an upshot of the meaning of the word 'metaphysics'. Its basic propositions can never be taken from experience, nor can its basic concepts; for it is not to be physical but metaphysical knowledge, so it must lie beyond experience. (Kant 1997, §1)

Besides the conceptual analysis of the term 'metaphysics', another motivating ground for an aprioristic conception that Kant appealed to was his conviction that metaphysics should yield knowledge that is necessarily true. But metaphysical claims that are based on empirical sources or premises, Kant holds, cannot be established as holding with necessity.

That metaphysics is an a priori discipline was also a shared assessment throughout much of the twentieth century. In their influential dismissal of metaphysics as being non-verifiable and nonsensical, the logical empiricists assumed the priori character of metaphysics throughout. Kant's view that metaphysics is by definition non-empirical persisted.[1] Thus, even after the demise of verificationist arguments against the meaningfulness of metaphysical claims and despite criticisms of the a priori–a posteriori distinction, the assumption that metaphysics needs to be a priori was widely taken for granted. When metaphysics resurged in the analytical tradition, metaphysicians such as E. J. Lowe argued that metaphysics is an a priori enterprise in virtue of the fact that it is concerned with essences:

> [M]etaphysics is most perspicuously characterized as the science of essence – a primarily a priori discipline concerned with revealing, through rational reflection and argument, the essences of entities, both actual and possible, with a view to articulating the fundamental structure of reality as a whole. (Lowe 2011, 99–100)

Similarly, Kit Fine characterised metaphysics, among other features, by its aprioricity. Metaphysics, Fine writes,

> should be concerned with the nature of reality; it should operate at a high level of generality; its method of enquiry should be a priori and its means of expression transparent; and it should be capable of providing a foundation for all other enquiry into the nature of reality. (Fine 2012, 24)

The fact that metaphysics has often been characterised in terms of a priori sources does not mean that this characterisation was perceived as being descriptively adequate with respect to how metaphysics was in fact conducted. Kant, for instance, criticises the use of empirical premises by some authors in his discussion of the cosmological and the physico-teleological argument for the existence of God (Kant 1999, 569–83). So, the apriority of metaphysics was put forward – at least in his case – as a normative ideal.

Such an ideal puts restrictions on the methods that are admissible in metaphysics. (I take a metaphysical method to be a procedure that allows us to establish a metaphysical claim.[2]) Conceptual analysis, appeal to intuitions or transcendental arguments are all methods that potentially establish metaphysical claims without appeal to empirical premises and might arguably yield necessary truths. Methods that appeal to the natural

---

[1] Thus, one author, Arthur Pap, explicitly states in his *Elements of Analytic Philosophy*:

> [...] it would seem that in its traditional usage the word has at least one clear negative meaning: metaphysics is not an empirical inquiry. To speak of an experimental verification of a metaphysical statement sounds absurd: if the statement is experimentally verifiable, then it is by definition empirical, not metaphysical. (Pap 1949, 7)

[2] See Tahko (2015), Nolan (2016) and Schrenk (2016) for recent discussions of methods in analytic metaphysics.

sciences, e.g., to established theories or to features of scientific practice, are, however, not admissible given the ideal of an aprioristic metaphysics.

The most distinctive feature of the approach I have taken in this book is that it takes the success of certain features of scientific practice as an explanandum of an inference to the best explanation. Because the relevant features of scientific practice are empirically accessible only, the approach I have taken is opposed to purely aprioristic conceptions of metaphysics. It is opposed to a Kantian conception of metaphysics for at least two reasons. First, inference to the best explanation is a non-deductive form of inference and thus does not suffice to establish conclusions that hold with certainty or epistemic necessity (unless they are transformed into transcendental arguments). Second, even though inference to the best explanation might (in principle) concern non-empirical explananda, e.g., certain features of mathematical objects, the features of scientific practice that we looked at in the previous chapters are clearly empirical phenomena.

8.2   Inductive Metaphysics

By the middle of the nineteenth century a purely aprioristic conception of metaphysics as elaborated in the idealistic systems of Hegel, Fichte and others was perceived as having failed (Beiser 2014, 188). In retrospect, various responses to this state of affairs can be distinguished, among them positivism and Neo-Kantianism. Both positivists and Neo-Kantians were critical of metaphysics and were subsequently perceived as 'enemies of metaphysics' (Külpe 1898, 25).

From the perspective of a minimal metaphysics of scientific practice, another movement that evolved as a response to the demise of aprioristic metaphysics is more interesting, a movement that was later dubbed 'Inductive Metaphysics'[3]. On the one hand, inductive metaphysicians – in contrast to positivists and Neo-Kantians – insisted on there being genuine metaphysical issues that deserve to be tackled, and on the other hand – in contrast to aprioristic conceptions of metaphysics – they held not only that metaphysics is science-informed but, furthermore, that it uses the same inductive methods as the sciences.

Erich Becher, looking back on the development of this movement, praised Gustav Theodor Fechner as the founding father of inductive

---

[3] My brief characterisation is based on a more detailed account of the development and the program of inductive metaphysics in Scholz (2018). Gustav Theodor Fechner, Rudolf Hermann Lotze, Eduard von Hartmann, Oswald Külpe, Wilhelm Wundt, Hans Driesch and Erich Becher are some of the authors who are usually discussed as inductive metaphysicians or as their precursors.

metaphysics: 'Fechner introduced into recent metaphysics the empirical-inductive method, i.e., a research procedure (*Forschungsverfahren*) that starts from but in its inferences goes beyond experience' (Becher, quoted in Scholz 2018, my translation). Oswald Külpe was probably the first to use the term 'Inductive Metaphysics' explicitly. In 1898 he wrote, targeting positivist and neo-Kantian critics of metaphysics:

> Kant was wrong when he assumed that metaphysics is possible only as an a priori science (*Wissenschaft*) based on pure reason and untenable once this procedure is exposed as vain and problematic. The idea of an inductive metaphysics that grows out of the other sciences and complements them was neither conceived by him nor is it touched by his criticisms. (Külpe 1898, 25, my translation)

One important issue for inductive metaphysicians – at least in hindsight – is the question of how far metaphysics is allowed to grow out of or go beyond experience. Hans Driesch, to illustrate this problem, argued for a neo-vitalist position positing non-mechanical entelechies to account for biological processes. Can the posit of entelechies be justified within a metaphysics that relies on inductive methods?[4] Even though there was some discussion of how inductive methods work in the sciences and how they might be applied in metaphysics – for instance, in the works of Wilhelm Wundt (see Seide, manuscript) – there was not much concern about *constraints* regarding inferences that go beyond experience. This led – even within inductive metaphysics – to what was perceived as philosophical speculation. Oswald Külpe, for example, accused one of his fellow inductive metaphysicians, Eduard von Hartmann, for this very reason of producing 'conceptual poetry'.[5]

Given that the minimal metaphysics of scientific practice as developed in the previous chapters is an example of an inductive metaphysics, it has to face the question of what constraints, if any, there are when it comes to inference to the best explanation in metaphysics.

8.3   Inference to the Best Explanation as an Inductive Method in Metaphysics

In an inference to the best explanation 'one infers, from the premise that a given hypothesis would provide a "better" explanation for the

---

[4] Weber argues that Driesch's arguments fail to be convincing even from the perspective of his contemporaries because these arguments rely too much on non-empirical and not-argued-for assumptions (Weber 1999).
[5] A term used by F. A. Lange (and later by Carnap) to characterise all of metaphysics.

evidence than would any other hypothesis, to the conclusion that the given hypothesis is true' (Harman 1965, 89). The fate of inference to the best explanation was closely tied to that of scientific realism. As long as in twentieth-century philosophy of science scientific theories were taken to be mere instruments for prediction, the question of whether the best explanation allows one to infer to the truth of the hypothesis does not arise. Within Logical Empiricism the debate about scientific realism was considered to be a pseudo-problem, or, as Nagel put it with respect to realism and instrumentalism about theories, '[. . .] the opposition between these views is a conflict over preferred modes of speech' (Nagel 1961, 152). It was in the context of the threat of incommensurability that the question of whether theoretical terms refer to objects was reassessed. According to Putnam's 'negative argument for scientific realism', an instrumentalist understanding of theories is responsible for the problem of incommensurability (Putnam 1975, 197–8). Once a realistic understanding of scientific theories had been accepted by many philosophers of science, typical deliberations and inferences in the sciences had to be understood as inferences to the best explanation, i.e., as inferences to the *truth* of hypotheses. Moreover, on a meta-level, the very debate over scientific realism was itself framed as the issue over whether the success of scientific theories is best explained by the truth of scientific realism. Thus, the role of inference to the best explanation in the sciences and in the context of the scientific realism debate became a widely discussed issue (for an overview see Lipton 2004).

Let us return to the issue of inference to the best explanation in metaphysics. While in the last quarter of the twentieth century conceptual analysis was perhaps the most salient method metaphysicians self-consciously adopted (e.g., Chalmers 1996 or Jackson 1998), more recently the view that there is a methodological continuity between the sciences and metaphysics has become more prominent. Sider, Hawthorne and Zimmerman, for instance, hold that 'Metaphysicians use standards for choosing theories that are like the standards used by scientists (simplicity, comprehensiveness, elegance, and so on)' (Sider, Hawthorne and Zimmerman 2007, 6; see also Paul 2012, 9 and Schrenk 2016, 288 for a list of further quotes). In this context, inference to the best explanation appears to be as problematic or unproblematic a tool for the metaphysician as it is for the scientist. As a matter of fact, during the heyday of conceptual analysis, Armstrong, in arguing for his account of laws of nature in terms of a necessitation relation, already employed this form of inference.[6]

8.3.1 Challenges to Inference to the Best Explanation in Metaphysics

In what follows I will not rehearse the entire debate about the role of inference to the best explanation. Rather, I assume that inference to the best explanation is a legitimate method in the sciences. My focus will be on some recent arguments to the effect that there are important disanalogies between the sciences, on the one hand, and metaphysics, on the other, such that inference to the best explanation, while a reliable method in the former, fails to be a reliable method in the latter.

Let me start with van Fraassen. As a dissenter with respect to scientific realism, van Fraassen had raised some objections against inference to the best explanation in the sciences early on (van Fraassen 1980, 19–40). More important for the question under consideration are his more specific objections in the context of his criticism of analytic metaphysics and ontology:

> the very phrase 'inference to the best explanation' should wave a red flag for us. What is good, better, best? What values are slipped in here, under a common name, and where do they come from? [...] If simplicity, strength, coherence, and all those other explanatory virtues are to be placed in the balance against truth, may we please be shown the balance, the gauge, the units, the scale? As long as we have nothing like that, the defence of ontology as scientific points only to the form its theory and theory choice take. But nonsense can come in the same form as wisdom. Form is nowhere near enough. (van Fraassen 2002, 14–16)

Merely using the same words in evaluating metaphysical hypotheses does not ensure that the same criteria are used as in scientific theory choice. Ladyman adds the consideration that prediction and empirical adequacy play a different role in inferences to the best explanations in physics and metaphysics:

> Explanatory power plays the role it does in theory choice because of the relationship between theoretical explanation and the empirical virtues of scientific theories. We have inductive grounds for believing that pursuing simplicity and explanatory power in science will lead to empirical success, but no such grounds where we are dealing with distinctively metaphysical

---

[6] 'The postulation of that extra thing is a case of inference to the best explanation. It is rational to postulate what best explains the phenomena' (Armstrong 1983, 55).

explanations, since the latter is completely decoupled from empirical success. (Ladyman 2012, 46)

Two issues may be distinguished. First, Ladyman suggests that empirical adequacy is not only a criterion on its own (which is conspicuously absent from the list of criteria that Sider, Hawthorne and Zimmerman identified as being relevant for both scientists and metaphysicians) but might furthermore be relevant for an explication of other criteria, such as simplicity or elegance. Second, while there are inductive reasons to believe that inferences to the best explanation are truth-conducive in the sciences, there is no such evidence for the case of metaphysics.[7]

Wrapping things up, there are a number of challenges for a minimal metaphysics of scientific practice, given that it relies on inference to the best explanation as its distinctive method.[8]

- What are the criteria that tell us whether an explanation in metaphysics is the best explanation?
- How are these criteria to be spelled out?
- What is the role of experience and prediction with respect to these criteria and to inference to the best explanation in metaphysics?
- Do we have reasons to believe that inference to the best explanation is truth-conducive in metaphysics?

Answering these questions in full generality would surely require another book. What I will do in the next section is to look at some of the specific cases in which I relied on inference to the best explanation and explicate how these questions can be answered in these particular contexts. In Section 8.3.3 I will briefly look at the role of experience in the context of the particular inference to the best explanation I used in the previous chapters.

---

[7] This point has been forcefully taken up by Beebee:

> In other words, to the extent that – and only to the extent that – best explanations in science have turned out to be predictively superior to their rivals, we have inductive grounds in the scientific case for thinking that inference to the best explanation is truth-conducive. But no such inductive grounds exist in the case of metaphysics. (Beebee 2018, 7)

[8] Strictly speaking, I should distinguish two ways of talking about inference to the best explanation that I have not distinguished so far: (a) inference to the best explanation as a particular form of an inference and (b) inference to the best explanation as a method (see Scholz 2015). Inference to the best explanation as a method is a procedure that might comprise various steps, e.g., explicating the empirical phenomenon that constitutes the intended explanandum, constructing metaphysical hypotheses that offer explanations of these phenomena, deriving further consequences from these hypotheses and testing their plausibility on the basis of their fit with accepted background knowledge, etc.

8.3.2 Spelling out 'Best' in Some Inferences to the Best Explanations in Metaphysics

As already indicated, I will not provide general answers to the questions just listed. In fact, it seems that we have good reasons to believe that there are no such general answers. For instance, there is no reason to assume that inference to the best explanation in general relativity, solid-state physics or evolutionary biology works in exactly the same way. Even if explanations in these fields may appeal to simplicity, coherence or other theoretical virtues, these will be spelt out differently in different contexts. The fact that what counts as the best explanation is highly context dependent is no problem as long as, within these contexts, what is best can be explicated with sufficient clarity. For the same reason, there is no need to specify what 'best explanation' means for all of metaphysics. Against van Fraassen's request to deal with the first two items of the preceding list of challenges, I thus use what van Fraassen in his pragmatic account of explanation called the 'legitimate rejection' of a presupposition (van Fraassen 1980, 146) of the challenges, namely, that an adequate answer gives us a completely general account of 'better' or 'best' and of criteria such as simplicity.

Rather, I intend to recapitulate some of the cases in which I have used 'best explanation' and to state clearly the criteria that were pertinent and why, therefore, inference to the best explanation is a reliable procedure in these particular contexts.

The first case is a case of an *inference to the only available explanation.* (By 'available' I mean 'currently available'.) Before I go into the details, I should briefly say something about the notion of explanation I appeal to when I claim that a metaphysical hypothesis accounts for an explanandum. I don't want to commit myself to a specific notion of what explanation consists in. For convenience, I assume that for an explanans E to explain an explanandum e, it has to be shown how e depends on E, such that given E, e obtains. However, an account according to which for E to explain e, it has to be the case that, given E, e is to be expected should work as well.

The particular case I want to look at is the argument for dispositions that was discussed in the context of the extrapolation argument. To recapitulate a passage from Section 2.4.2:

> In my terminology, Cartwright argues that we need dispositions to understand why extrapolation works. When we consider Galileo's law of free fall ('nothing else is going on, disturbances are absent') to be relevant for

a falling body in a medium (different context, 'mixed circumstances'), we assume that something carries over from the first (ideal) situation to the second situation: a property that is manifest in the first situation but fails to be manifest in the second (though, of course, it is instantiated in both situations).

The (empirical) explanandum in this case is the fact that we sometimes successfully extrapolate from one situation to a qualitatively different situation. Extrapolation works; that is, on the basis of extrapolations we can successfully predict and manipulate the behaviour of systems. The explanation accounts for this explanandum by postulating a dispositional property that is present in both situations. The fact that we are dealing with the same property accounts for why the one situation is relevant for the other; the fact that the property may or may not be manifest accounts for the fact that the situations may nevertheless be qualitatively different.

So, what we have here is the explanation of an empirical fact – the success of the method of extrapolation. It is, furthermore, the best available explanation of the success of extrapolation simply because it is the only one currently available. This assessment may, of course, change over time. People may come up with competing explanations or with new empirical evidence that might be deemed relevant for an explanation of extrapolation, for instance, hitherto ignored features of scientific practice. But that is a situation which we have in the sciences too. Thus, as it stands, the dispositional account is the best explanation we have. The fact that the dispositional account explains the success of extrapolation, the fact there are no plausible competing hypotheses, and the fact that there is no evidence or background knowledge that undermines the dispositional account provide reasons for tentatively accepting the hypothesis as true.

The situation in the case just outlined is markedly different from another case that was also discussed in the context of the extrapolation argument, the attempt to provide a metaphysical explanation of 'same property' in terms of either a universal or perfectly similar tropes. In this latter case, I argued, there is no evidence from the success of scientific practice that would make a difference for these hypotheses. So, in this latter case, as opposed to the case of the dispositional properties, we should remain agnostic.

The second case I want to look at is an inference to the best explanation where we have a competition of different explanatory hypotheses. One important presupposition for such a competition to get off the ground is a consensus about the explanandum. If there is no such consensus, there is no point in asking for the best explanation. It may indeed be a hitherto overlooked disanalogy between the sciences and metaphysics that it is more difficult in metaphysics to establish such a consensus. In physics, the subcommunities of researchers who establish phenomena as explananda and those who develop theoretical hypotheses to explain them are typically different. By contrast, in metaphysics, for example in the debate about Humeanism, the very same people who advance the metaphysical explanation try to delineate the explanandum. It may thus be no surprise that no consensus emerges about whether, e.g., the Humean mosaic needs to be explained or should be taken as primitive.

Be that as it may, in the case I will focus on here there is no problem of a missing consensus concerning the explanandum. The empirical phenomenon to be explained is the reductive practice described in Chapter 5. In Chapters 6 and 7 we discussed several metaphysical hypotheses that provide explanations for this practice. One particular explanandum was the supervenience of macroscopic phenomena on microscopic phenomena (not a feature of scientific practice, but to be accepted as an intermediate explanatory hypothesis, if it is accepted that reductive practices aim at reductions in terms of very general theories). There is no change in macro-properties without a change in micro-properties. All the accounts discussed (Foundationalism, Physical Eliminativism and Ontologically Neutral Monism) provide explanations for supervenience, as I argued in Chapters 6 and 7, i.e., given the truth of any of these hypotheses, it becomes clear how supervenience depends on the explanans and why it is thus to be expected that the macro supervenes on the micro. However, in the case at hand, I argued in Chapter 7 that Ontologically Neutral Monism provides the *best* explanation and that we should take this to be evidence for the truth of Ontologically Neutral Monism. Why is that so?

The reason is that in the case at hand, the various hypotheses are more or less parsimonious with respect to the assumptions they make about the structure of reality. Parsimony, in this context, concerns not the number of different kinds of entities that are assumed to exist but, rather, the strength of assumptions about reality. For reasons to be explained later, a minimal metaphysics of scientific practice considers that hypothesis which makes the weakest assumptions about reality to be the best explanation.

In the case at hand, the different metaphysical hypotheses are fairly transparently related to each other in terms of what they positively assume about the structure of reality. When confronted with cases of supervenience, as, for example, in thermodynamics and statistical mechanics, Ontological Neutral Monism ('H_I') does not assume there to be two sets of properties, for instance, microscopic or fundamental physical facts, on the one hand, and macroscopic facts, on the other. To explain supervenience it suffices to assume that there is one kind of fact/entity/event (*'onekindoffact'*) and allow there to be various theoretical accounts of such facts/entities/events that are reductively related, such that for every change described in macroscopic terms there is a change described in microscopic terms.

Physical Eliminativism ('H_2') agrees with Ontologically Neutral Monism that there is one kind of facts/entities/events but, furthermore, claims that the facts/entities/events in question are fundamental-physical facts. To explain supervenience it suffices to assume that there are these fundamental-physical facts (*'funphys'*). Even though the hypotheses do not differ with respect to parsimony if parsimony is spelled out in terms of how many different kinds of facts/entities/events are postulated, Ontologically Neutral Monism is more parsimonious in its assumptions about these facts/entities/events because it does not commit itself to the claim that the facts/entities/events in question are of a fundamental-physical nature.

Physical Foundationalism ('H_3') agrees with Physical Eliminativism that there are fundamental-physical facts, but it furthermore asserts that there are additional layers of facts that are ontologically posterior with respect to the fundamental-physical facts (*'addition'*). Supervenience is then explained in terms of this dependence structure.

In the case we are considering, we have a hypothesis H_1 that explains our explanandum e, where the explanatory work is done by *onekindoffact*. There are, furthermore, two rival hypotheses, H_2 and H_3, that explain e as well. Because *onekindoffact* on its own is more parsimonious than the conjunction of *onekindoffact* with either *funphys* or *funphys* and *addition*, we have a clear case of H_1 being the most parsimonious explanation of e – the explanation that makes the weakest assumptions. This allows us, furthermore, to say in which sense H_1 provides the *best* explanation. While H_1 explains e in terms of *onekindoffact* only, H_2 and H_3 postulate additional facts that are explanatorily irrelevant.

The evidence we gain from being able to explain e (supervenience) is thus evidence for *onekindoffact* only. It does not provide evidence for *funphys* or for *addition*. In the context under consideration, *funphys* and *addition* are just two arbitrary claims that have been added to *onekindoffact*, which does all the explanatory work. The hypotheses *funphys* and *addition* are what is called in confirmation theory 'irrelevant conjuncts'. We do have positive evidence for *onekindoffact*, and that is the only claim we have positive evidence for. Since evidence for a hypothesis is evidence for the truth of the hypothesis, it makes perfectly good sense for a metaphysics of scientific practice to go for minimality, i.e., to accept those hypotheses as true for which we have positive evidence and to remain agnostic with respect to everything else.[9] A *minimal* metaphysics of scientific practice is a metaphysics that (1) accepts only hypotheses for which we have evidence (evidence for the truth of the hypotheses) in the form of explanation of features of scientific practice and that (2) refrains from postulating any structure that is explanatorily irrelevant for understanding the success of the scientific practice we have or other phenomena for which we have good reasons to assume that they provide truth-conducive evidence.[10]

To come back to van Fraassen's challenges: There may be no general explication of what is 'best' in inferences to the best explanations in metaphysics. However, there are special cases in which we are able to specify the criteria for best explanations with sufficient clarity and where we have reason to assume that the evidence we gain from successful inferences to the best explanations is truth-conducive.

8.3.3   Inference to the Best Explanation in Metaphysics and the Role of Experience

In the case of a minimal metaphysics for scientific practice, experience plays an important role in inferences to the best explanation because in many explanatory arguments the explananda are empirically accessible features of scientific practice. Metaphysical hypotheses may or may not be able to account for such empirical phenomena. If they are able to account for the phenomena, they may be better or worse in a sense that can be clearly articulated in at least some cases. In these cases, we have good reasons to assume that inference to the best explanation is as truth-conducive as it is in everyday contexts or in the sciences.

But what about the alleged disanalogy according to which there is inductive evidence for the success of inference to the best explanation in the sciences (and in everyday contexts), while there is no such evidence for inference to the best explanation in metaphysics? Ladyman and Beebee, who stress this point, do not cite any studies – though Ladyman mentions a few cases in which inference to the best explanation has led to empirically

---

[9] Of course, the situation will in general be much more complicated. The evidence may be weaker or stronger, there may be counter-evidence, there will be background assumptions, etc. I leave aside those complications.

[10] This applies to the case of causal intuitions – as opposed to other intuitions typically appealed to in metaphysics (see Chapter 4).

successful theories (in the sense of being predictively superior to their rivals). The evidence, however, is at best anecdotal. And the evidence people advance in the context of pessimistic meta-induction may provide anecdotal evidence for the claim that inference to the best explanation in the sciences sometimes goes wrong. It seems to me that an inductive argument of this kind is in need of more empirical support.

Finally, it may be a valid objection that metaphysics fails to be predictively successful because it does not make any empirical predictions to start with. Still, by relying on inferences to the best explanation, a minimal metaphysics of scientific practice is – to some extent at least – epistemically risky with respect to empirical phenomena because there may be features of scientific practice that have been overlooked or not yet analysed and may lead to the need to revise the original hypotheses. It seems to me that inferences to the best explanation in a minimal metaphysics of scientific practice are very much in the same boat as those in the natural sciences that concern purely historical phenomena, as is the case in some areas of astrophysics or evolutionary biology.

8.4 Naturalised Metaphysics

Is the minimal metaphysics of scientific practice an example of naturalised metaphysics? That depends on what one means by 'naturalised metaphysics'. On a very liberal reading, for a metaphysical theory to be classified as 'naturalised metaphysics' it suffices for a metaphysical account to be 'science-informed' or 'science-inspired'. A minimal metaphysics of scientific practice would surely be a naturalised metaphysics in this sense. But then almost any view will qualify as naturalised metaphysics. Kant's *Metaphysical Foundations of Natural Science* (Kant 2004) or Schelling's *Ideas for a Philosophy of Nature* (Schelling 1988) would certainly be good examples.

A less liberal reading of naturalised metaphysics starts with the fact that the term has been coined in analogy with naturalised epistemology. Quine, in his 'Epistemology Naturalized' suggests that what he considers to be one of the main epistemological issues, the relation between evidence and our description of the world, should be considered as an empirical problem. As a consequence, '[e]pistemology, or something like it, simply falls into place as a chapter of psychology and hence of natural science' (Quine 1969, 82). By analogy, just as naturalised epistemology considers epistemology to be a part of psychology, naturalised metaphysics intends metaphysics to become a chapter of the natural sciences. This less liberal reading seems to be what has been advocated by some authors, such as Kincaid, Ladyman and Ross. Kincaid, for instance, characterises naturalised (scientific) metaphysics in the following terms: '[...] it is only by means of scientific results and scientific methods that metaphysical knowledge is possible' (Kincaid 2013, 3).

In its more radical form, scientific or naturalistic metaphysics does not only advocate 'an extreme scepticism about metaphysics when it is based on conceptual analysis tested against intuition, and about any alleged apriori truths that such intuitions and analyses might yield' (Kincaid 2013, 3). It furthermore holds that metaphysical questions that either cannot be directly answered by the sciences or are not in the service of the sciences should have no place in metaphysics: 'If metaphysics is to be part of the pursuit of objective knowledge, it must be integrated with science. Genuinely naturalized metaphysics must go beyond mere consistency with current science; it must be directly motivated by and in the service of science.' (Ladyman and Ross 2013, 109).

While the minimal metaphysics of scientific practice advocated in the earlier chapters shares the scepticism against metaphysics purely based on conceptual analysis and intuitions (though it allows for exceptions if we have good reasons to hold that the intuitions in question are truth-conducive), it does not agree with radical naturalised metaphysics that the only purpose of metaphysics is to be serviceable to the sciences. Why should it? Questions about dispositions and categorical properties or about universal and tropes are perfectly fine questions irrespective of whether or not physics or any other science has any use for the accounts developed in answering these questions. It may be difficult to find evidence for some of the views developed in these contexts, but that is a different issue.

A minimal metaphysics of scientific practice takes metaphysical issues seriously whether or not they promote natural science. The purpose of the foregoing chapters is to show that with respect to some of these issues progress can be made if the right kind of evidence is taken into account.

Bibliography

Albert, D. (2000). *Time and Chance*. Cambridge, Mass.: Harvard University Press.

Albert, D. (2015). *After Physics*. Cambridge, Mass.: Harvard University Press.

Aristotle. (1984). Metaphysics. In J. Barnes, ed., *The Complete Works of Aristotle, Vol. II*. Princeton: Princeton University Press.

Armstrong, D. M. (1983). *What Is a Law of Nature?* Cambridge: Cambridge University Press.

Armstrong, D. M. (2010). *Sketch for a Systematic Metaphysics*. Oxford: Oxford University Press.

Ashcroft, N. W. and Mermin, N. D. (1976). *Solid State Physics*. Philadelphia: Saunders College.

Barnes, E. (2018). Symmetric Dependence. In R. Bliss and G. Priest, eds., *Reality and Its Structure: Essays in Fundamentality*. Oxford: Oxford University Press, pp. 50–69.

Batterman, R. W. (2011). Emergence, Singularities, and Symmetry Breaking. *Foundations of Physics*, 41, 1031–50.

Baumgartner, M. (2010). Interventionism and Epiphenomenalism. *Canadian Journal of Philosophy*, 40, 359–83.

Beatty, J. (1995). The Evolutionary Contingency Thesis. In G. Wolters and J. G. Lennox, eds., *Concepts, Theories, and Rationality in the Biological Sciences, The Second Pittsburgh-Konstanz Colloquium in the Philosophy of Science*. Pittsburgh: University of Pittsburgh Press.

Bechtel, W. and Richardson, R. (1993). *Discovering Complexity: Decomposition and Localization as Strategies in Scientific Research*. Princeton: Princeton University Press.

Beebee, H. (2018). Philosophical Scepticism and the Aims of Philosophy. *Proceedings of the Aristotelian Society*, 118, 1–24.

Beiser, F. (2014). *The Genesis of Neo-Kantianism 1796–1880*. Oxford: Oxford University Press.

Bennett, K. (2008). Exclusion Again. In J. Hohwy and J. Kallestrup, eds., *Being Reduced*. Oxford: Oxford University Press, pp. 280–305.

Bird, A. (2005). The Dispositionalist Conception of Laws. *Foundations of Science*, 10, 353–70.

Bird, A. (2007). *Nature's Metaphysics: Laws and Properties*. Oxford: Oxford University Press.

Blanchard, T. and Schaffer, J. (2017). Cause without Default. In H. Beebee, C. Hitchcock and H. Price, eds., *Making a Difference*. Oxford: Oxford University Press, pp. 175–214.

Bliss, R. (2019). What Work the Fundamental? *Erkenntnis*, 84, 359–79.

Block, N. (2003). Do Causal Powers Drain Away? *Philosophy and Phenomenological Research*, 67, 133–50.

Bohm, A. (1986). *Quantum Mechanics: Foundations and Applications*. New York: Springer.

Breuer, T. (1995). The Impossibility of Accurate State Self-Measurements. *Philosophy of Science*, 62 (2), 197–214.

Broad, C. D. (1925). *Mind and Its Place in Nature*. London: Routledge and Kegan Paul.

Butterfield, J. (2011). Less Is Different: Emergence and Reduction Reconciled. *Foundations of Physics*, 41, 1065–135.

Campbell, N. A. and Reece, J. B. (2002). *Biology*. 6th edn. San Francisco: Benjamin Cummings.

Carrier, M. (1998). In Defense of Psychological Laws. *International Studies in the Philosophy of Science*, 12, 217–32.

Cartwright, N. (1983). *How the Laws of Physics Lie*. Oxford: Oxford University Press.

Cartwright, N. (1989). *Nature's Capacities and Their Measurement*. Cambridge: Cambridge University Press.

Cartwright, N. (2007). *Hunting Causes and Using Them*. Cambridge: Cambridge University Press.

Cartwright, N. (2008), Reply to Stathis Psillos. In S. Hartmann, C. Hoefer and L. Bovens, eds., *Nancy Cartwright's Philosophy of Science*. London: Routledge.

Cartwright, N. and Hardie, J. (2012). *Evidence-Based Policy*. Oxford: Oxford University Press.

Castellani, E. (2003). Symmetry and Equivalence. In K. Brading and E. Castellani, eds., *Symmetries in Physics*. Cambridge: Cambridge University Press, pp. 425–36.

Castellani, E. and de Haro, S. (2020). Duality, Fundamentality, and Emergence. In D. Glick, G. Darby and A. Marmodoro, eds., *The Foundation of Reality: Fundamentality, Space and Time*. Oxford: Oxford University Press.

Chakravartty, A. (2007) *A Metaphysics for Scientific Realism*. Cambridge: Cambridge University Press, pp. 195–216.

Chakravartty, A. (2013). Review of S. Mumford and R. L. Anjum, 'Getting Causes from Powers'. *British Journal for the Philosophy of Science*, 64, 895–9.

Chakravartty, A. (2017). *Scientific Ontology*. Oxford: Oxford University Press.

Chalmers, D. (1996). *The Conscious Mind*. Oxford: Oxford University Press.

Choi, S. and Fara, M. (2016). Dispositions. In E. N. Zalta, ed., *The Stanford Encyclopedia of Philosophy*. Spring 2016 edn. Available at: https://plato.stanford.edu/archives/spr2016/entries/dispositions/

Collins, J., Hall, N. and Paul, L. (2004). Introduction. In J. Collins, N. Hall and L. Paul, eds., *Causation and Counterfactuals*. Cambridge, MA: MIT Press, pp. 1–57.

Correia, F. (2008) Ontological Dependence. *Philosophy Compass*, 3, 1013–32.

Crowther, K. (2020). What Is the Point of Reduction in Science? *Erkenntnis*, 85, 1437–60.

Danks, D. (2009). The Psychology of Causal Perception and Reasoning. In H. Beebee, C. Hitchcock and P. Menzies, eds., *The Oxford Handbook of Causation*. Oxford: Oxford University Press, pp. 447–70.

Darrigol, O. and Renn, J. (2013). The Emergence of Statistical Mechanics. In J. Buchwald and R. Fox, eds., *The Oxford Handbook of the History of Physics*. Oxford: Oxford University Press, pp. 765–88.

Descartes, R. (1985). *The Philosophical Writings of Descartes, Vol.* I. Transl. J. Cottingham, R. Stoothoff and D. Murdoch. Cambridge: Cambridge University Press.

Descartes, R. (1991). *The Philosophical Writings of Descartes, Vol.* III. Transl. J. Cottingham, R. Stoothoff, D. Murdoch and A. Kenny. Cambridge: Cambridge University Press.

Dizadji-Bahmani, F., Frigg, R. and Hartmann, S. (2010). Who's Afraid of Nagelian Reduction? *Erkenntnis*, 73, 393–412.

Dobson, C. M. (2003). Protein Folding and Misfolding. *Nature*, 426, 884–90.

Dowe, P. (2000). *Physical Causation*. Cambridge: Cambridge University Press.

Dowe, P. (2009). Causal Process Theories. In H. Beebee, C. Hitchcock and P. Menzies, eds., *The Oxford Handbook of Causation*. Oxford: Oxford University Press, pp. 213–33.

Dresden, M. (1974). Reflections on 'Fundamentality and Complexity'. In Ch. Enz and J. Mehra, eds., *Physical Reality and Mathematical Description*. Dordrecht: Springer, pp. 133–66.

Earman, J. (1986). *A Primer on Determinism*. Dordrecht: D. Reidel.

Earman, J. (1989). *World Enough and Spacetime: Absolute versus Relational Theories of Space and Time*. Cambridge, Mass.: MIT Press.

Earman, J. and Roberts, J. (1999). Ceteris Paribus, There Is No Problem of Provisos. *Synthese*, 118, 439–78.

Earman, J., Roberts, J. and Smith, S. (2002). Ceteris paribus lost. In J. Earman et al., eds., Ceteris paribus laws. *Erkenntnis*, 52 (Special issue), 281–301.

Eddington, A. (1964). *The Nature of the Physical World* (Everyman's Library). London: Dent.

Ehlers, J. (1986). On Limit Relations between and Approximate Explanations of Physical Theories. In B. Marcus, G. J. W. Dorn and P. Weingartner, eds., *Logic, Methodology and Philosophy of Science, VII*. Amsterdam: Elsevier, pp. 387–403.

Ehlers, J. (1997). Examples of Newtonian limits of relativistic spacetimes. *Classical and Quantum Gravity*, 14: A119–A126.

Einstein, A. (1921). Geometry and Experience. In A. Einstein [1983]. *Sidelights on Relativity*. New York: Dover. pp. 27–56.

Einstein, A. (1979) *Autobiographical Notes*. A Centennial Edition, ed. by P. A. Schilpp, Chicago: Open Court Publishing Company.

Ellis, B. (2001). *Scientific Essentialism*. Cambridge: Cambridge University Press.

Feyerabend, P. K. (1962). Explanation, Reduction and Empiricism. In H. Feigl and G. Maxwell, eds., *Scientific Explanation, Space, and Time*. Minneapolis: University of Minnesota Press, pp. 28–97.

Field, H. (2003). Causation in a Physical World. In M. Loux and D. Zimmerman, eds., *The Oxford Handbook of Metaphysics*. Oxford: Oxford University Press, pp. 435–60.

Fine, K. (1994). Essence and Modality. *Philosophical Perspectives*, 8, 1–16.

Fine, K. (1995). The Logic of Essence. *Journal of Philosophical Logic*, 24, 241–73.

Fine, K. (2012). What Is Metaphysics? In T. Tahko, ed., *Contemporary Aristotelian Metaphysics*. Cambridge: Cambridge University Press, pp. 8–25.

Fischer, F. (2018). *Natural Laws as Dispositions*. Berlin: de Gruyter.

Fodor, J. (1974). Special Sciences, or the Disunity of Science as a Working Hypothesis. *Synthese*, 28, 97–115.

Fodor, J. (1991). You Can Fool Some People All of the Time, Everything Else Being Equal; Hedged Laws and Psychological Explanations. *Mind*, 100, 19–34.

French, S. (2014). *The Structure of the World*. Oxford: Oxford University Press.

Friedman, M. (1986). *Foundations of Space-Time Theories: Relativistic Physics and Philosophy of Science*. Princeton: Princeton University Press.

Frisch, M. (2014). *Causal Reasoning in Physics*. Cambridge: Cambridge University Press.

Galilei, G. (1954). *Dialogues Concerning Two New Sciences*. Transl. H. Crew and A. de Salvio. New York: Dover Publications.

Gaukroger, S. (2002). *Descartes' System of Natural Philosophy*. Cambridge: Cambridge University Press.

Godfrey-Smith, P. (2009). Causal Pluralism. In H. Beebee, C. Hitchcock and P. Menzies, eds., *The Oxford Handbook of Causation*. Oxford: Oxford University Press, pp. 326–37.

Hall, N. (2004). Two Concepts of Causation. In J. Collins, N. Hall and L. Paul, eds., *Causation and Counterfactuals*. Cambridge MA: MIT Press, pp. 225–76.

Hall, N. (2015). Humean Reductionism about Laws of Nature. In B. Loewer and J. Schaffer, eds., *The Blackwell Companion to David Lewis*. Oxford: Blackwell, pp. 262–77.

Halpern, J. Y. (2015). A Modification of the Halpern-Pearl Definition of Causality. In Q. Yang and M. Wooldridge, eds., *Proceedings of the 24th International Joint Conference on Artificial Intelligence (IJCAI 2015)*, Palo Alto: AAAI Press, pp. 3022–33.

Halpern, J. Y. and Pearl, J. (2005). Causes and Explanations: A Structural-Model Approach. Part I: Causes. *The British Journal for Philosophy of Science*, 56, 843–87.

Hart, H. L. and Honoré, A. M. (1959). *Causation in the Law*. Oxford: Oxford University Press.

Havas, P. (1974). Causality and Relativistic Dynamics. In H. Wolfe and W. Rolnick, eds., *AIP Conference Proceedings*, 16, College Park: AIP Publishing, pp. 23–47.

Hawley, K. (2006). Science as a Guide to Metaphysics? *Synthese*, 149 (3), 451–70.

Healey, R. (2013). Physical Composition. *Studies in History and Philosophy of Modern Physics*, 44, 48–62.

Hempel, C. (1988). Provisoes: A Problem concerning the Inferential Function of Scientific Theories. *Erkenntnis*, 28, 147–64.

Hicks, M. T. (2018) Dynamic Humeanism. *British Journal for the Philosophy of Science*, 69 (4), 983–1007.

Hitchcock, C. (2001). The Intransitivity of Causation Revealed in Equations and Graphs. *Journal of Philosophy*, 98, 273–99.

Hitchcock, C. (2016). Probabilistic Causation. In E. N. Zalta, ed., *The Stanford Encyclopedia of Philosophy*. Winter 2016 edn. Available at: https://plato .stanford.edu/archives/win2016/entries/causation-probabilistic/.

Hitchcock, C. and Knobe, J. (2009). Cause and Norm. *Journal of Philosophy*, 106, 587–612.

Hitchcock, C. and Woodward, J. (2003). Explanatory Generalizations, Part II, Plumbing Explanatory Depth. *Nous*, 37, 181–99.

Hoefer, C. and Smeenk, C. (2016). Philosophy of the Physical Sciences. In P. Humphreys, ed., *Oxford Handbook of the Philosophy of Science*. Oxford: Oxford University Press, pp. 115–36.

Hoffmann-Kolss, V. (2014). Interventionism and Higher-Level Causation. *International Studies in the Philosophy of Science*, 28 (1), 49–64.

Hofweber, T. (2009). Ambitious, Yet Modest, Metaphysics. In D. Chalmers, D. Manley and R. Wasserman, eds., *Metametaphysics*. Oxford: Oxford University Press, pp. 260–89.

Howard, D. and Stachel, J. (2000). *Einstein: The Formative Years, 1879–1909* (Einstein Studies, vol. 8). Boston: Birkhäuser.

Hoyningen-Huene, P. (1993). *Reconstructing Scientific Revolutions – Thomas S. Kuhn's Philosophy of Science*. Chicago: The University of Chicago Press.

Hoyningen-Huene, P. (1994). Emergenz versus Reduktion. In G. Meggle and U. Wessels, eds., *Analyomen 1*. Berlin: de Gruyter, pp. 324–32.

Humphreys, P. (1997). How Properties Emerge. *Philosophy of Science*, 64, 1–17.

Hüttemann, A. (1998). Laws and Dispositions. *Philosophy of Science*, 65, 121–35.

Hüttemann, A. (2004). *What's Wrong with Microphysicalism?* London: Routledge.

Hüttemann, A. (2005). Explanation, Emergence and Quantum-Entanglement. *Philosophy of Science*, 72, 114–27.

Hüttemann, A. (2014). Ceteris Paribus Laws in Physics. *Erkenntnis*, 79, 1715–28.

Hüttemann, A. (2015). Physicalism and the Part-Whole-Relation. In T. Bigaj and C. Wüthrich, eds., *Metaphysics in Contemporary Physics*. Poznan Studies in the Philosophy of the Sciences and the Humanities 104, pp. 32–44.

Hüttemann, A. and Love, A. C. (2011). Aspects of Reductive Explanation in Biological Science: Intrinsicality, Fundamentality, and Temporality. *The British Journal for Philosophy of Science*, 62, 519–49.

Hüttemann, A. and Love, A. C. (2016). Reduction. In P. Humphreys, ed., *The Oxford Handbook of Philosophy of Science*. Oxford: Oxford University Press, pp. 460–84.

Hüttemann, A. and Papineau, D. (2005). Physicalism Decomposed. *Analysis*, 65, 33–9.

Hüttemann, A., Kühn, R. and Terzidis, O. (2015). Stability, Emergence and Part-Whole-Reduction. In B. Falkenburg and M. Morrison, eds., *Why More Is Different: Philosophical Issues in Condensed Matter Physics and Complex Systems*. Dordrecht: Springer, pp. 169–200.

Ismael, J. (2015). How to Be Humean. In B. Loewer and J. Schaffer, eds., *The Blackwell Companion to David Lewis*. Oxford: Blackwell, pp. 188–205.

Jaag, S. and Loew, C. (2020). Making Best Systems Best for Us. *Synthese*, 197, 2525–2550.

Jackson, F. (1998). *From Metaphysics to Ethics*. Oxford: Oxford University Press.

Johansson, I. (1980). Ceteris Paribus Clauses, Closure Clauses and Falsifiability. *Journal for the General Philosophy of Science*, 10, 16–22.

Joos, E., Zeh, H. D., Kiefer, C., et al. (2003). *Decoherence and the Appearance of a Classical World in Quantum Theory*. 2nd edn. New York: Springer.

Kahneman, D. and Miller, D. T. (1986). Norm Theory: Comparing Reality to Its Alternatives. *Psychological Review*, 93, 136–53.

Kant, I. (1997). *Prolegomena to Any Future Metaphysics*, ed. by G. Hatfield. Cambridge: Cambridge University Press.

Kant, I. (1999). *Critique of Pure Reason*, ed. by P. Guyer and A. Wood. Cambridge: Cambridge University Press.

Kant, I. (2004). *Metaphysical Foundations of Natural Science*, ed. by M. Friedman. Cambridge: Cambridge University Press.

Kemeny, J. G. and Oppenheim, P. (1956). On Reduction. *Philosophical Studies*, 7, 6–19.

Kennedy, J. B. (1995). On the Empirical Foundations of the Quantum No-signalling Proofs. *Philosophy of Science*, 62, 543–60.

Ketterle, W. (2007). Bose-Einstein Condensation: Identity Crisis for Indistinguishable Particles. In J. Evans and A. S. Thorndike, eds., *Quantum Mechanics at the Crossroads – New Perspectives from History, Philosophy and Physics*. Heidelberg: Springer, pp. 159–82.

Kim, J. (1984). Epiphenomenal and Supervenient Causation. *Midwest Studies in Philosophy*, 9, 257–70. Reprinted in J. Kim, ed. (1993) *Supervenience and Mind*. Cambridge: Cambridge University Press, pp. 92–108.

Kim, J. (1985). Psychological Laws. In E. LePore and B. McLaughlin, eds., *Actions and Events*, Oxford: Oxford University Press, pp. 369–86.

Kim, J. (1988). Explanatory Realism, Causal Realism, and Explanatory Exclusion. *Midwest Studies of Philosophy*, 12, 225–39. Reprinted in J. Kim (2010). *Essays in the Metaphysics of Mind*. Oxford: Oxford University Press, pp. 148–66.

Kim, J. (1994). Explanatory Knowledge and Metaphysical Dependence. *Philosophical Issues* 5, 51–69. Reprinted in J. Kim (2010). *Essays in the Metaphysics of Mind*. Oxford: Oxford University Press, pp. 167–86.

Kim, J. (2002). Responses. *Philosophy and Phenomenological Research*, 65, 671–80.

Kim, J. (2005). *Physicalism or Something Near Enough*. Princeton: Princeton University Press.

Kincaid, H. (2004). Are There Laws in the Social Sciences?: Yes. In C. Hitchcock, ed., *Contemporary Debates in the Philosophy of Science*. Oxford: Blackwell, pp. 168–86.

Kincaid, H. (2013). Introduction: Pursuing a Naturalist Metaphysics. In D. Ross, J. Ladyman and H. Kincaid, eds., *Scientific Metaphysics*. Oxford: Oxford University Press, pp. 1–26.

Kirchhoff, G. (1876). *Vorlesungen über mathematische Physik: Mechanik*. Leipzig: Teubner.

Kistler, M. (2012). Powerful Properties and the Causal Basis of Dispositions. In A. Bird, B. Ellis and H. Sankey, eds., *Properties, Powers and Structures. Issues in the Metaphysics of Realism*. Oxford: Routledge, pp. 119–37.

Kuhn, T. (1996). *The Structure of Scientific Revolutions*. 3rd edn. Chicago: The University of Chicago Press.

Külpe, O. (1898). *Einleitung in die Philosophie*. 2nd ed., Leipzig: Hirzel.

Kutach, D. (2011). The Asymmetry of Influence. In C. Callender, ed., *The Oxford Handbook of Philosophy of Time*. Oxford: Oxford University Press, pp. 247–75.

Ladyman, J. (2012). Science, Metaphysics and Method. *Philosophical Studies*, 160, 31–51.

Ladyman, J. and Ross, D. (2007). *Every Thing Must Go. Metaphysics Naturalized*. Oxford: Oxford University Press.

Ladyman, J. and Ross, D. (2013). The World in the Data. In H. Kincaid, J. Ladyman and D. Ross, eds., *Scientific Metaphysics*. Oxford: Oxford University Press, pp. 108–50.

Landsman, N. P. (2007). Between Classical and Quantum. In J. Earman and J. Butterfield, eds., *Handbook of the Philosophy of Physics*. Amsterdam: Elsevier, pp. 417–553.

Lange, M. (1993). Natural Laws and the Problem of Provisos. *Erkenntnis*, 38, 233–48.

Lange, M. (2009). *Laws and Lawmakers: Science, Metaphysics, and the Laws of Nature*. Oxford: Oxford University Press.

Lepore, E. and Loewer, B. (1987). Mind Matters. *Journal of Philosophy*, 93, 630–42.

Lewis, D. (1986). Causation, Postscript to 'Causation'. In *Philosophical Papers, Vol. II*, Oxford: Oxford University Press, pp. 159–213.

Lipton, P. (1999). All Else Being Equal. *Philosophy*, 74, 155–68.

Lipton, P. (2004). *Inference to the Best Explanation*. 2nd edn. London: Routledge.

Livanios, V. (2010). Symmetries, Dispositions and Essences. *Philosophical Studies*, 148, 295–305.

Loewer, B. (2002). Comments on Jaegwon Kim's Mind and the Physical World. *Philosophy and Phenomenological Research*, 65, 655–62.

Lowe, E. J. (2011). The Rationality of Metaphysics. *Synthese*, 178 (1), 99–109.

Mach, E. (1872). *History and Root of the Principle of the Conservation of Energy*. Transl. P. E. B. Jourdain (1911). Chicago: The Open Court Publishing.

Mach, E. (1900). *Principien der Wärmelehre. Principles of the Theory of Heat: Historically and Critically Elucidated*. Transl. T. J. McCormack (1986). Dordrecht: D. Reidel.

Mach, E. (1982). *Die Mechanik*. Darmstadt: Wissenschaftl. Buchgesellschaft.

Mackie, J. (1965). Causes and Conditions. *American Philosophical Quarterly*, 2, 245–64.

Mackie, J. (1980). *The Cement of the Universe*, Oxford: Oxford University Press.

Maudlin, T. (1998). Part and Whole in Quantum Mechanics. In E. Castellani, ed., *Interpreting Bodies*. Princeton: Princeton University Press, pp. 46–60.

Maudlin, T. (2003). Distilling Metaphysics from Quantum Physics. In M. Loux and D. Zimmerman, eds., *Oxford Handbook of Metaphysics*. Oxford: Oxford University Press, pp. 461–87.

Maudlin, T. (2004). Causation, Counterfactuals and the Third Factor. In J. Collins, N. Hall and L. Paul (eds), *Causation and Counterfactuals*. Cambridge MA: MIT Press, pp. 419–43.

Maudlin, T. (2007). *The Metaphysics within Physics*. Oxford: Clarendon Press.

McKenzie, K. (2011). Arguing against Fundamentality. *Studies in the History and Philosophy of Modern Physics*, 42, 244–55.

McKenzie, K. (2019). Fundamentality. In S. Gibb, R. Hendry and T. Lancaster, eds., *The Routledge Handbook of Emergence*. London: Routledge, pp. 54–64.

McKitrick, J., Marmodoro, A., Mumford, S., et al. (2013). Causes as Powers. *Metascience*, 22, 545–59.

McLaughlin, B. and Bennett, K. (2018). Supervenience. In E. Zalta, ed., *The Stanford Encyclopedia of Philosophy*. Winter 2018 edn. Available at: https://plato.stanford.edu/archives/win2018/entries/supervenience/.

Menon, T. and Callender, C. (2013). Turn and Face the Strange . . . Ch-ch-changes: Philosophical Questions Raised by Phase Transitions. In R. Batterman, ed., *The Oxford Handbook of Philosophy of Physics*. Oxford: Oxford University Press, pp. 189–223.

Merricks, T. (2001). *Objects and Persons*. Oxford, New York: Clarendon Press; Oxford University Press.

Mill, J. S. (1836). On the Definition and Method of Political Economy. In D. Hausman, ed. (2008). *The Philosophy of Economics. An Anthology*. 3rd edn. New York: Cambridge University Press, pp. 41–58.

Mill, J. S. (1974). *A System of Logic. Ratiocinative and Inductive*, Vol. VII and VIII of J. S. Mill. *Collected Works*, Toronto: University of Toronto Press.

Mitchell, S. (2002). Ceteris Paribus – An Inadequate Representation of Biological Contingency. *Erkenntnis*, 52, 329–50.

Mitchell, S. (2003). *Biological Complexity and Integrative Pluralism*. Cambridge: Cambridge University Press.

Morganti, M. (2015). Dependence, Justification and Explanation: Must Reality Be Well-Founded? *Erkenntnis*, 80, 555–72.

Mumford, S. (1998). *Dispositions*. Oxford: Oxford University Press.

Mumford, S. (2014). Contemporary Efficient Causation: Aristotelian Themes. In T. Schmaltz, ed., *Efficient Causation: A History*. Oxford: Oxford University Press, pp. 317–39.

Mumford, S. and Anjum, R. L. (2011). *Getting Causes from Powers*. Oxford: Oxford University Press.

Nagel, E. (1961). *The Structure of Science: Problems in the Logic of Scientific Explanation*. New York: Harcourt, Brace and World.

Newton, I. (1999). *The Principia*. Transl. I. B. Cohen and A. Whitman. Berkeley: University of California Press.

Ney, A. (2012). Neo-Positivist Metaphysics. *Philosophical Studies*, 160 (1), 53–78.

Nickles, T. (1973). Two Concepts of Intertheoretic Reduction. *Journal of Philosophy*, 70, 181–201.

Nobel Prize Press Release (2001). Available at: www.nobelprize.org/prizes/physics/2001/press-release/.

Nolan, D. (2016). Method in Analytic Metaphysics. In H. Cappelen, T. Szabó Gendler and J. Hawthorne, eds., *The Oxford Handbook of Philosophical Methodology*, Oxford: Oxford University Press, pp. 159–78.

North, J. (2013). The Structure of a Quantum World. In D. Albert and A. Ney, eds., *The Wave Function: Essays on the Metaphysics of Quantum Mechanics*. Oxford: Oxford University Press, pp. 184–202.

Norton, J. (2012). Approximation and Idealization: Why the Difference Matters. *Philosophy of Science*, 79, 207–32.

Oppenheim, P. and Putnam, H. (1958). Unity of Science as a Working Hypothesis. In H. Feigl, M. Scriven and G. Maxwell, eds., *Concepts, Theories, and the Mind-Body Problem*. Minneapolis: University of Minnesota Press, pp. 3–36.

Pap, A. (1949). *Elements of Analytical Philosophy*. New York: Macmillan.

Paul, L. (2012). Metaphysics as Modeling: The Handmaiden's Tale. *Philosophical Studies*, 160 (1), 1–29.

Paul, L. and Hall, N. (2013). *Causation: A User's Guide*. Oxford: Oxford University Press.

Pearl, J. (2000). *Causality*. Cambridge: Cambridge University Press.

Pietroski, P. and Rey, G. (1995). When Other Things Aren't Equal: Saving Ceteris Paribus Laws from Vacuity. *British Journal for the Philosophy of Science*, 46, 81–110.

Popper, K. (1959). *The Logic of Scientific Discovery*. London: Hutchinson.

Popper, K. (1974). Replies to My Critics. In P. A. Schilpp, ed., *The Philosophy of Karl Popper*. La Salle: Open Court Press.

Psillos, S. (2002). *Causation and Explanation*. Montreal: McGill-Queen's University Press.

Putnam, H. (1975). Explanation and Reference. In H. Putnam, *Mind, Language and Reality, Philosophical Papers Vol. 2*, Cambridge: Cambridge University Press, pp. 196–214.

Quine, W. (1969). Epistemology Naturalized. In *Ontological Relativity and Other Essays*, New York: Columbia University Press, pp. 69–90.

Reutlinger, A. (2011). A Theory of Non-Universal Laws. *International Studies in Philosophy of Science*, 25, 97–117.

Reutlinger, A. and Saatsi, J., eds. (2018) *Explanation beyond Causation*. Oxford: Oxford University Press.

Reutlinger, A., Schurz, G., Hüttemann, A. and Jaag, S. (2019). Ceteris Paribus Laws. In E. N. Zalta, ed., *The Stanford Encyclopedia of Philosophy*. Spring 2017 edn. Available at: https://plato.stanford.edu/archives/spr2017/entries/ceteris-paribus/.

Roberts, J. (2004). There Are No Laws in the Social Sciences. In C. Hitchcock, ed., *Contemporary Debates in the Philosophy of Science*. Oxford: Blackwell, pp. 168–85.

Rohrlich, F. (1988). Pluralistic Ontology and Theory Reduction in the Physical Sciences. *The British Journal for the Philosophy of Science*, 39, 295–312.

Rohrlich, F. and Hardin, L., (1983). Established Theories. *Philosophy of Science*, 50, 603–17.

Rosen, G. (2010). Metaphysical Dependence: Grounding and Reduction. In B. Hale and A. Hoffmann, eds., *Modality: Metaphysics, Logic, and Epistemology*. Oxford: Oxford University Press, pp. 109–36.

Ruben, D. (1990). *Explaining Explanation*. London: Routledge.

Russell, B. (1912/13). On the Notion of Cause. *Proceedings of the Aristotelian Society*, 13, 1–26.

Salmon, W. (1984). *Scientific Explanation and the Causal Structure of the World*. Princeton: Princeton University Press.

Sarkar, S. (1998). *Genetics and Reductionism*. Cambridge: Cambridge University Press.

Schaffer, J. (2004). Causes Need Not Be Physically Connected to Their Effects: The Case for Negative Causation. In C. Hitchcock, ed., *Contemporary Debates in Philosophy of Science*. London: Blackwell, pp. 197–216.

Schaffer, J. (2005). Contrastive Causation. *Philosophical Review*, 114 (3), 327–58.

Schaffer, J. (2010). Monism: The Priority of the Whole. *Philosophical Review*, 119 (1), 31–76.

Schaffner, K. F. (1967). Approaches to Reduction. *Philosophy of Science*, 34, 137–47.

Schaffner, K. F. (1969). The Watson-Crick Model and Reductionism. *The British Journal for the Philosophy of Science*, 20, 325–48.

Schaffner, K. F. (1976). Reductionism in Biology: Prospects and Problems. In R. Cohen, C. Hooker, A. Michalos and J. Van Evra, eds., *Proceedings of the 1974 Biennial Meeting of the Philosophy of Science Association*. Dordrecht: Reidel, pp. 613–32.

Schaffner, K. F. (1993). *Discovery and Explanation in Biology and Medicine*. Chicago: University of Chicago Press.

Scheibe, E. (1991a). Predication and Physical Law. *Topoi*, 10, 3–12.

Scheibe, E. (1991b). General Laws of Nature and the Uniqueness of the Universe. In E. Aggazzi and A. Cordero, eds., *Philosophy and the Origin and the Evolution of the Universe*. Kluwer: Dordrecht, pp. 341–60.

Scheibe, E. (1999). *Die Reduktion Physikalischer Theorien: Teil II, Inkommensurabilität und Grenzfallreduktion*. Berlin: Springer.

Scheibe, E. (2006). *Die Philosophie der Physiker*. Munich: Beck.

Schelling, F. (1988). *Ideas for a Philosophy of Nature*. Cambridge: Cambridge University Press.

Schlosshauer, M. (2004). Decoherence, the Measurement Problem, and Interpretations of Quantum Mechanics. *Review of Modern Physics*, 76, 1267–305.

Scholz, O. (2015). Texte interpretieren – Daten, Hypothesen und Methoden. In J. Borkowski, S. Descher, F. Ferder and P. Heine, eds., *Literatur interpretieren: Interdisziplinäre Beiträge zu Theorie und Praxis*. Münster: Mentis, pp. 147–71.

Scholz, O. (2018). Induktive Metaphysik – ein vergessenes Kapitel der Metaphysikgeschichte. In D. Hommen and D. Sölch, eds., *Philosophische Sprache zwischen Tradition und Innovation (Festschrift for Christoph Kann)*, Frankfurt am Main: Peter Lang, pp. 267–89.

Schrenk, M. (2007a). Can Capacities Rescue Us from Ceteris Paribus Laws? In M. Kistler and B. Gnassounou, eds., *Dispositions and Causal Powers*. Aldershot: Ashgate, pp. 221–47.

Schrenk, M. (2007b). *The Metaphysics of Ceteris Paribus Laws*. Heusenstamm: Ontos.

Schrenk, M. (2010). On the Powerlessness of Necessity. *Nous*, 44 (4), 725–39.

Schrenk, M. (2011). Interfering with Nomological Necessity. *The Philosophical Quarterly*, 61, 577–97.

Schrenk, M. (2016) *Metaphysics of Science*. London: Routledge.

Schurz, G. (2001). Causal Asymmetry, Independent versus Dependent Variables and the Direction of Time. In W. Spohn, M. Ledwig and M. Esfeld, eds., *Current Issues in Causation*. Paderborn: Mentis, pp. 47–67.

Schurz, G. (2002). Ceteris Paribus Laws: Classification and Deconstruction. *Erkenntnis*, 52, 351–72.

Schurz, G. and Gebharter, A. (2016). Causality as a Theoretical Concept: Explanatory Warrant and Empirical Content of the Theory of Causal Nets. *Synthese*, 193, 1073–103.

Schuster, J. (2013). Cartesian Physics. In J. Buchwald and R. Fox, eds., *Oxford Handbook of the History of Physics*. Oxford: Oxford University Press, pp. 56–95.

Seide, A. (manuscript). Wilhelm Wundts *Logik* als Auftakt zu einer induktiven Metaphysik.

Shoemaker, S. (1980). Causality and Properties. In P. van Inwagen, ed., *Time and Cause: Essays Presented to Richard Taylor*. Dordrecht: D. Reidel, pp. 109–35.

Sider, T., Hawthorne, J. and Zimmerman, D. (2007). *Contemporary Debates in Metaphysics*. Oxford: Blackwell.

Sklar, L. (1993). *Physics and Chance: Philosophical Issues in the Foundations of Statistical Mechanics*. New York: Cambridge University Press.

Skow, B. (2016). *Reasons Why*. Oxford: Oxford University Press.

Spirtes, P., Glymour, C. and Scheines, R. (2000). *Causation, Prediction, and Search*. Cambridge, Mass.: MIT Press.

Stanford, P. K. (2006). *Exceeding Our Grasp*. Oxford: Oxford University Press.

Steel, D. (2008). *Across the Boundaries*. Oxford: Oxford University Press.

Strawson, P. (1959). *Individuals*. London: Routledge.

Suárez, F. (1866). *Disputationes Metaphysicae*. In M. André and C. Berton, eds., *Opera Omnia*, Vol. 25 and 26, Paris, reprint 1998. Hildesheim: Olms.

Suppes, P. (1970). A Probabilistic Theory of Causality. *Acta Philosophica Fennica*, XXIV, 1–130.

Tahko, T. (2015). *An Introduction to Metametaphysics*. Cambridge: Cambridge University Press.

Tahko, T. (2018). Fundamentality. In E. Zalta, ed., *The Stanford Encyclopedia of Philosophy*. Fall 2018 edn. Available at: https://plato.stanford.edu/archives/fal l2018/entries/fundamentality/.

Tan, P. (2019). Counterpossible Non-Vacuity in Scientific Practice. *Journal of Philosophy*, 116, 32–60.

Trogdon, K. (2018). Inheritance Arguments for Fundamentality. In R. Bliss and G. Priest, eds., *Reality and Its Structure: Essays in Fundamentality*. Oxford: Oxford University Press, pp. 182–98.

Tugby, M. (2013). Platonic Dispositionalism. *Mind*, 122, 451–80.

Uffink, J. (2007). Compendium of the Foundations of Classical Statistical Physics. In J. Earman and J. Butterfield, eds., *Handbook of the Philosophy of Physics*. Amsterdam: Elsevier, pp. 923–1074.

Van Fraassen, B. (1980). *The Scientific Image*. Oxford: Oxford University Press.

Van Fraassen, B. (1989). *Laws and Symmetry*. Oxford: Oxford University Press.

Van Fraassen, B. (2002). *The Empirical Stance*. New Haven: Yale University Press.

Van Inwagen, P. (1990). *Material Beings*. Ithaca: Cornell University Press.

Vetter, B. (2015). *Potentiality. From Dispositions to Modality*. Oxford: Oxford University Press.

Weber, M. (1999). Hans Drieschs Argumente für den Vitalismus. *Philosophia Naturalis*, 36 (2), 263–93.

Williamson, J. (2009). Probabilistic Theories. In H. Beebee, C. Hitchcock and P. Menzies, eds., *The Oxford Handbook of Causation*. Oxford: Oxford University Press, pp. 185–212.

Wilson, J. (2012). Fundamental Determinables. *Philosophers' Imprint*, 12 (4), 1–17.

Wilson, J. (2014). No Work for a Theory of Grounding. *Inquiry*, 57, 535–79.

Wimsatt, W. C. (1976). Reductive Explanation: A Functional Account. In R. S. Cohen, ed., *Proceedings of the Philosophy of Science Association-1974*, pp. 671–710.

Woodward, J. (1992). Realism about Laws. *Erkenntnis*, 36, 181–218.

Woodward, J. (2002). There Is No Such Thing as a Ceteris Paribus Law. *Erkenntnis*, 57, 303–28.

Woodward, J. (2003). *Making Things Happen*. Oxford: Oxford University Press.

Woodward, J. (2015). Interventionism and Causal Exclusion. *Philosophy and Phenomenological Research*, 91 (2), 303–47.

Woodward, J. (2018). Laws: An Invariance-based Account. In W. Ott and L. Patton, eds., *Laws of Nature*. Oxford: Oxford University Press, pp. 158–80.

Woodward, J. and Hitchcock, C. (2003). Explanatory Generalizations, Part I, A Counterfactual Account. *Nous*, 37, 1–24.

Zeh, H.-D. (2001). *The Physical Basis of the Direction of Time*. 4th edn. Heidelberg: Springer.

Index

a priori metaphysics. *see* metaphysics, a priori
Albert, David, 106
Anjum, Rani Lill, 74–8
anomaly, 141–9
Aristotle, 83
Armstrong, David, 23, 60–2, 207

backing relation, 131, 165–8, 174–5, 195
Batterman, Robert, 150
Becher, Erich, 205
Beebee, Helen, 214
Bennett, Karen, 191
Bird, Alexander, 23, 37, 40, 50, 61, 74–6, 80
Blanchard, Thomas, 95, 115
Bliss, Ricki, 161
Bose-Einstein-condensate, 21–2
Boyle-Mariotte law. *see* ideal gas law
bridge laws, 133
Broad, Charles Dunbar, 142, 154

Callender, Craig, 149, 150
capacities. *see* dispositions
carbon monoxide, 55, 169
Carrier, Martin, 40
Cartwright, Nancy, 41, 46, 50–2, 58, 83, 177, 211
categorical property, 47–50, 53, 58, 62, 78
causal modelling, 128
causal pluralism, 83
causal processes, 99, 191, *see* processes
causation
    asymmetry of, 86, 103, 106, 166
    closed-system, 104, 107
    conserved quantities theory, 99, 119, 191, 193
    counterfactual theory, 100, 114, 193
    definition, 89
    definition (augmented), 117–18
    dispositional theory of. *see* dispositions and
        causation
    disruptive concept of, 92, 99, 101, 102, 105–7,
        117, 128
    focal concept of, 84, 87, 101

interventionist theory, 126, 193
    negative, 119–21
    process theory of, 100, 110, 113, 118, 119, 191, 193,
        see also causation, conserved quantity theory
    reductive account of, 97
    regularity theory of, 85, 100, 106, 191
    transitivity of, 110, 125
ceteris paribus law, 38–81, 92, 114
    confirmation problem, 44, 47, 51, 67
    exclusive, 40, 41, 46, 65, 92, 114
    Lange's dilemma, 42, 43, 44, 65, 66
    semantic problem, 44, 47, 51, 65
Chakravartty, Anjan, 50, 52, 78, 100
conceptual analysis, 108, 203, 204, 207
conservation laws, 80, 111
Correia, Fabrice, 164
counterfactual dependence, 110, 114, 126, 127, 191
counterfactuals, 16, 26, 64, 112–14, 126

decoherence, 151–2, 156, 201–2
default relativity, 95, 115
dependence
    counterfactual. *see* counterfactual dependence
    generic existential, 163–4, 167, 195
    modal, 107, 160–2
    mutual, 102, 167, 171–4, 178, 181
    nomological, 107
    non-modal, 160–2, 164
    rigid existential, 164
Descartes, René, 5, 141
    cartesian physics, 141–2
determinism, 101
disconnection, problem of, 108, 119
dispositional modality, 23, 37, 76
dispositions
    and causation, 82, 90, 92–4, 98–100
    characterisation of, 47–50
    dispositional account of laws, 38–81
    multi-track, 46, 47–50, 58–63, 78–80,
        92–3, 163
double prevention, 119

Dowe, Phil, 98, 108, 110, 119
Dresden, Max, 129–30

Earman, John, 42–5, 54, 57, 71–2
Eddington, Arthur, 197–8, 201
Eddington's table, 197–8
Einstein, Albert, 14, 201
Eliminativism, Physical, 130–1, 157, 186–93, 194,
    195–7, 198, 212–13
Ellis, Brian, 50, 78
emergence, 142–3, 145, 153, 154
    of classicality, 151–2, 156
entropy, 144, 201
essences, 37, 50, 73, 78, 80, 81, 204
exclusion argument, 191, *see also*
    overdetermination argument
explanation
    causal, 82, 84, 86, 87, 96, 109, 157, 165, 166–7,
        175, 188
    part-whole, 55–7, 163, 164, 165–6, 167–75, 186,
        195, 197, 199
    standard, 18–19, 35
explanatory asymmetry, 158, 165, 166–7, 174–5
explanatory realism, 165, 166, 195
extrapolation, 50–65
    argument from, 64, 75, 78, 81, 210–11
    problem of, 50–1

Fechner, Gustav Theodor, 85, 205
Feyerabend, Paul, 134, 135
Fine, Kit, 160, 161, 164, 204
Fischer, Florian, 52
Fodor, Jerry, 40
Foundationalism, 131, 157, 159–63, 168, 173–4,
    175–7, 184, 185, 186, 187, 198, 212, 213
French, Steven, 5, 22, 80, 198
Frisch, Mathias, 98
fundamentality, 5, 19, 129, 162, 182

Galilei, Galileo, 33
Galileo's law, 13, 15, 18, 31, 35, 39, 41, 51, 52, 54, 59,
    91, 92, 210
Gebharter, Alexander, 126
generalisation, 13–19, 98, 107, 114, 140
    internal vs. external, 15–19, 24–31, 35–6, 59, 73
generality, 142, 143, 175–7, 193
Giere, Ronald, 14
grounding, 161, 196

Hall, Ned, 24, 83, 111, 116, 125, 127
Halpern, John, 113
Harman, Gilbert, 207
Havas, Peter, 102, 106
Hawley, Katherine, 3
Healey, Richard, 163

Hempel, Carl, 19, 42
Hicks, Michael T., 24, 28
Hitchcock, Christopher, 16, 17, 19, 25, 26,
    125, 126
Hoefer, Carl, 130, 182–5
Hoffmann-Kolss, Vera, 193
Hooke's law, 24
Hoyningen-Huene, Paul, 141, 143
Humeanism, 6, 24, 28, 37, 78, 81, 152, 162, 212
Humphreys, Paul, 154
hyperintensionality, 162

ideal gas law, 17, 27–32, 63
indistinguishability of particles, 21–2
inductive metaphysics. *see* metaphysics, inductive
inference to the best explanation, 4, 11, 37, 52, 53,
    177, 196, 205, 209, 214–15
instrumentalism about theories, 207
interference
    characterisation of, 97, 99
    relevant vs. irrelevant, 110–13
interventionism. *see* causation, interventionist
    theory
intrinsic properties, 20–3, 74, 94, 95, 145, 201
invariance, 24, 26, 28–36, 37, 39, 72–81, 107,
    163, 175
    of law equation, 28–37
Ismael, Jenann, 28

Jaag, Siegfried, 24, 28

Kant, Immanuel, 203, 204, 215
Kim, Jaegwon, 165, 191, 192, 193
Kincaid, Harold, 40, 216
Kirchhoff, Gustav, 84
Kistler, Max, 59
Kuhn, Thomas, 141, 144, 145
Külpe, Oswald, 205–6
Kutach, David, 106

Ladyman, James, 3, 20–3, 208, 209, 214, 216
Lange, Friedrich Albert, 206
Lange, Marc, 30, 37, 41
law equation, 12, 16, 18, 23, 24, 27
law predicate, 14, 17, 18, 20, 23, 80
law statement
    dispositional characterisation, 64
    preliminary characterisation, 14
laws of nature, 11–12
    invariance account of, 36
    laws of composition, 55–63, 67, 77, 140, 142,
        153, 155, 169–75, 179–80, 195
Lewis, David, 109, 114, 152
Lipton, Peter, 65, 207
Livanios, Vassilios, 80

Loew, Christian, 24, 28
Loewer, Barry, 191
Logical Empiricism, 132, 134, 138, 203, 207
Loschmidt, Josef, 144
Lotka–Volterra equation, 15, 16, 20, 91, 92, 94

Mach, Ernst, 84–6, 87, 102, 105–7, 187
Mackie, John, 190
manifestation of a disposition, 47–50, 54, 58, 66,
    68, 75–80, 100
  complete vs. partial, 58, 62–3, 93
Maudlin, Tim, 2, 5, 97, 152
McKenzie, Kerry, 185
McKitrick, Jennifer, 100
Menon, Tarun, 149, 150
Merricks, Trenton, 189–93
metaphysics
  a priori, 203–5
  inductive, 205–6
  minimal metaphysics of scientific practice, 1,
    214–15
  naturalised, 215–16
micro-macro coherence, 141, 146,
    148–58, 168
  characterisation of, 145
Mill, John Stuart, 46, 54, 57, 58, 81, 85,
    104, 122
misconnection, problem of, 100, 110,
    113, 118
Mitchell, Sandra, 30, 40
modal surface structure, 12, 23, 36, 37,
    73, 76
Mumford, Stephen, 37, 47, 66, 74–8, 100

Nagel, Ernest, 132–8, 207
natural necessity, 12, 23, 33, 78
Neo-Kantianism, 205
Newton's first law, 39, 43, 68, 70, 76, 88, 90, 92,
    93, 98, 99
Newton's second law, 14, 24, 31, 32, 34, 39, 43,
    69, 92
Newtonian mechanics, 25, 54, 134, 137
Ney, Alyssa, 3, 5
Nickles, Thomas, 135–6, 137
nomological necessity, 23, 27–31, 33, 34, 37, 39,
    73, 175
nomological possibility, 24–6, 27, 72
North, Jill, 5

Ohm's law, 166, 174
omissions. *see* causation, negative
ontological priority, 130, 131, 160, 161–3, 184, 185,
    186, 193, 194
  and composition, 163–4
  and part-whole explanation, 164–75

and supervenience, 175–7
Ontologically Neutral Monism, 186–7, 195–7,
    198, 212–13
Oppenheim, Paul, 132, 137
overdetermination argument, 188–93

Paul, Laurie, 3, 109, 111, 116, 125, 127, 207
phase transition, 31, 149–50, 199–200
Pietroski, Paul, 44–6, 67, 69, 70, 71
Popper, Carl, 27, 40
powers. *see* dispositions
pre-emption, 109–19
processes, 98–100
  quasi-inertial, 88, 90–2, 94, 101, 103, 107,
    109–28
properties
  identity of, 79–81
protein-folding problem, 144–5
Psillos, Stathis, 34, 35
Putnam, Hilary, 132, 207

quantum mechanics, 3, 4, 9, 17, 54, 55, 56,
    59, 103, 137, 143, 147, 151, 155, 168,
    169, 170, 177, 179, 180, 185, 198,
    201, 202
quasi-inertial behaviour. *see also* processes,
    quasi-inertial
  characterisation of, 90–2
Quine, Willard, 215

reduction
  limit-case, 134–8, 156, 183
  micro-, 132–3, 138
  Nagelian, 133–4, 136, 138, 156
redundancy range, 126–8
relativity
  agent-, 116
  general theory of, 3, 137
  of system individuation, 96
  special theory of, 134, 135, 183
Reutlinger, Alexander, 45, 149
Rey, George, 44–6, 67, 69, 70,
    71
Roberts, John, 40, 42–5, 54, 57, 71–2
Rosen, Gideon, 161
Ross, Don, 20–3, 216
Russell, Bertrand, 84–6, 105–7

Saatsi, Juha, 149
Salmon, Wesley, 19, 99, 119
Schaffer, Jonathan, 89, 94, 95, 115, 119,
    160, 161
Scheibe, Erhard, 15, 38, 103
Schelling, Friedrich, 215
Schmalhausen's law, 15

Schmoranzer, Sebastian, 118
Scholz, Oliver, 205, 206, 209
Schrenk, Markus, 61, 66, 69, 204, 207
Schrödinger-equation, 13, 14, 17, 20, 24, 27, 39, 41, 42, 43, 44, 56, 59, 60, 62, 64, 65, 156, 170
Schurz, Gerhard, 40, 92, 126, 178–9
scientific practice
  characterisation of, 3–4, 11
Scriven, Michael, 19
second law of thermodynamics, 106, 144, 147, 200, 201
Seide, Ansgar, 206
Shoemaker, Sidney, 49, 79, 81
Skow, Bradford, 18
Smeenk, Chris, 182–5
Spohn, Wolfgang, 71
Steel, Daniel, 51, 53
Strawson, Peter, 4
structural equations, 126–8
Suárez, Francisco, 190, 191
supervenience, 175–7, 193–4, 196, 212, 213

and emergence, 154
systems, 19–23

tendencies. *see* dispositions
Theseus' ship, 163–4
transcendental argument, 52, 204, 205
tropes, 53, 211, 216
Tugby, Matthew, 47, 48, 76

universals, 52, 53, 61, 211, 216

van Fraassen, Bas, 14, 38, 116, 208, 210
van Inwagen, Peter, 189
Vetter, Barbara, 48

well-foundedness, 161
Wilson, Jessica, 62

Zeh, Heinz, 106
Zermelo, Ernst, 144